SpringerBriefs in Water Science and Technology

SpringerBriefs in Water Science and Technology present concise summaries of cutting-edge research and practical applications. The series focuses on interdisciplinary research bridging between science, engineering applications and management aspects of water. Featuring compact volumes of 50 to 125 pages (approx. 20,000–70,000 words), the series covers a wide range of content from professional to academic such as:

- Literature reviews
- In-depth case studies
- Bridges between new research results
- Snapshots of hot and/or emerging topics

Topics covered are for example the movement, distribution and quality of freshwater; water resources; the quality and pollution of water and its influence on health; and the water industry including drinking water, wastewater, and desalination services and technologies.

Both solicited and unsolicited manuscripts are considered for publication in this series.

Neha Saxena • Md. Merajul Islam
Deepa Sharma

Water Pollution and Remediation

A global concern

 Springer

Neha Saxena (ORCID)
School of Basic Sciences & Technology
IIMT University
Meerut, Uttar Pradesh, India

Md. Merajul Islam
School of Basic Sciences & Technology
IIMT University
Meerut, Uttar Pradesh, India

Deepa Sharma
School of Basic Sciences & Technology
IIMT University
Meerut, Uttar Pradesh, India

ISSN 2194-7244   ISSN 2194-7252   (electronic)
SpringerBriefs in Water Science and Technology
ISBN 978-3-031-76300-7   ISBN 978-3-031-76301-4   (eBook)
https://doi.org/10.1007/978-3-031-76301-4

This Springer imprint is published by the registered company Springer Nature Switzerland AG
The registered company address is: Gewerbestrasse 11, 6330 Cham, Switzerland

If disposing of this product, please recycle the paper.

Preface

Today, there is much discussion about the dangers of water contamination to human health. Numerous environmental studies and projects are undertaken and accomplished every year. Despite this, most developing countries continue to struggle with the inefficient administration of their water supplies. The problem of contaminated water and the method of purifying it are both extremely challenging, and much of the research done on the topic is probably not up to the mark to cope with the current environmental issues. Henceforth, appropriate methods need to be developed and established to improve water quality. Through this book, the authors aim to provide a framework for understanding the causes, pollutants, and potential remediation for water contamination. The text provides up-to-date reviews of the latest research and practices for removing contaminants from water through green methods.

Meerut, India   Neha Saxena

Md. Merajul Islam

Deepa Sharma

Acknowledgment

Writing this book has been an incredible journey, one that would not have been possible without the support and love of those closest to me.

To our wonderful children, Anannya, Aleena, Mariyam, Devanshi, and Haniya, their endless curiosity and boundless energy have been a constant source of inspiration. Their patience during the long hours of writing and the joyful interruptions has both kept us grounded and reminded us of what truly matters.

A special thank you to dearest Hachiko, whose wagging tail and comforting presence kept permanent companion during the late-night writing sessions. Its silent companionship and occasional nudges for walks provided the perfect breaks.

Authors are also deeply grateful to their families and friends who offered encouragement and understanding throughout this process. Their belief kept us motivated even on the toughest days.

Lastly, to you, the reader, thank you for sharing in this journey. I hope this book brings you as much insights in reading as it brought us in writing.

Contents

Chapter 1
Water Pollution: A Threat to Ecosystem

Abstract The threat that is being placed on our water ecosystem as a direct consequence of accelerated industrialisation, which helps to drive urbanization, is becoming extremely severe, which in turn reduces the amount of fresh water that is readily
available. Water pollution poses a significant threat to marine life, as well as to
plants, people, and the environment, and it also has the effect of changing the balance of the environment. This chapter will present the most current survey on water
pollution, as well as a classification based on the sources of pollution and an examination of the adverse impacts that water pollutants have on the ecosystem. In the
end, a survey will be conducted on worldwide legislations for the pollutions, particularly water pollution legislation, as well as the effectiveness of such legislation.

Keywords Water pollution · Toxicology · Ecosystem · Pollution legislation ·
Marine life

1.1 Introduction to Water Pollution

Pollution refers to the act of introducing detrimental substances into the environment. Water pollution is the presence of substances that reduce the quality of freshwater or marine water. Water pollution refers to the act of introducing pollutants into
surface groundwater or bodies of water such as lakes, streams, rivers, estuaries, and
seas (Noor et al. 2023). This introduction of substances reaches a level where it
disrupts the beneficial use of water or the normal functioning of ecosystems
(Wang et al. 2024). Water pollution encompasses the discharge of various
substances, such as chemicals, waste, and microorganisms, into bodies of water.
Additionally, it may involve the emission of energy, such as radioactivity or heat,
into these water sources.

© The Author(s), under exclusive license to Springer Nature   1
Switzerland AG 2024
N. Saxena et al., *Water Pollution and Remediation*, SpringerBriefs in Water
Science and Technology, https://doi.org/10.1007/978-3-031-76301-4_1

Water contamination is a significant problem that necessitates the implementation of effective regulations and tools to monitor and execute initiatives aimed at finding solutions. The yearly volume of wastewater released is estimated to be approximately 1500 cubic kilometres (Nationen 2003). To address the additional increase in water demand resulting from population growth, industrial development, and agricultural expansion, it is necessary to focus on limiting the pollution of freshwater sources and enhancing the treatment of wastewater. In recent decades, several substances have been detected in the aquatic environment, classified as either anthropogenic or naturally occurring compounds. This has led to growing concerns about the state of the world's ecosystem and surface waters (Wang et al. 2024).

Water is a vital necessity for the sustenance of the human species. According to the UNESCO-published 2021 World Water Development Report (Richard 2021), freshwater use has increased globally by a factor of six over the past century and has been rising steadily at a rate of about 1% per year since the 1980s. The rise in water demand is posing significant difficulties for water quality. The processes of industrialization, agricultural production, and urbanisation have led to the deterioration and contamination of the environment, causing harm to vital water bodies such as rivers and oceans (Noor et al. 2023). This, in turn, has negative consequences for human health and the ability in order to accomplish social advancement that is sustainable (Wang et al. 2024; Xu et al. 2022). Globally, industries and municipalities discharge almost 80% of their wastewater into the environment untreated, causing devastating effects on ecosystems and human health. In the least developed countries, where there is a severe lack of facilities for sewage purification and hygiene, such a ratio is more common.

1.2   Causes of Water Contamination

The primary sources of water contamination include industrialization, agricultural practices, natural factors, and inadequate water supply and sewage treatment infrastructure. Primarily, the main culprit behind water pollution is the industrial sector. This sector encompasses many industries such as distillery, tannery, pulp and paper, textile, food, iron and steel, nuclear, and others. Industrial production can result in the emission of a wide range of harmful compounds, including both organic and inorganic molecules, as well as toxic solvents and volatile organic chemicals. Insufficiently treated releases of these wastes into aquatic environments will result in water pollution (Chowdhary et al. 2020). Arsenic, cadmium, and chromium are essential contaminants released in wastewater, and the industrial sector plays a large role in the emission of dangerous pollutants (Lin et al. 2022). Due to the rapid pace of urbanisation, there has been a gradual rise in the amount of wastewater generated by industrial activities (Wu et al. 2020). Furthermore, foreign direct investment has a significant impact on water contamination resulting from industrialization.

Foreign direct investment has a positive correlation with industrial water contamination in less developed countries (Lin et al. 2022). Furthermore, there is a strong correlation between water contamination and agriculture. Agricultural activities, specifically the use of pesticides, nitrogen fertilisers, and organic agricultural wastes, are major contributors to water contamination (RCEP 1979). Agricultural operations contribute to the pollution of water with substances such as nitrates, phosphate, pesticides, soil sediments, salts, and pathogens (Parris 2014). In addition, agriculture has significantly impaired the original condition of all freshwater systems. Unprocessed or partially processed wastewater is extensively used for irrigation in regions of developing countries that have limited water resources, such as China and India. However, the sewage's contamination can have detrimental effects on both the environment and human health. Using China as a case study, the disparity in the amount and quality of surface water resources has resulted in the prolonged utilisation of wastewater irrigation in certain regions of developing nations to fulfil the water requirements of agricultural activities. This has consequently caused significant contamination of agricultural land and food, posing risks to food safety and human health due to the presence of pesticide residues and heavy metal pollution (Lu et al. 2015).

Water contamination, particularly in the Loess Plateau, is a significant health concern due to the use of pesticides. The Longitudinal Survey data shows a correlation between pesticide use and medical disability index among those over 65 years of age. The Musi River in India shows a higher prevalence of illness in areas where wastewater is used for irrigation compared to families using regular water. Water contamination is interconnected with natural elements, such as trace elements from natural weathering processes and human activities. High levels of salt and salinity in river water indicate poor water quality (Xiao et al. 2019; Lin et al. 2022). Hexavalent chromium pollution is the most common form of water pollution in the middle region of the Loess Plateau. Water supply and sewage treatment facilities play a crucial role in determining drinking water quality, particularly in developing nations. Inadequate investment in basic water supply and treatment infrastructure has resulted in water contamination, increased occurrence of contagious and parasitic illnesses, and increased vulnerability to industrial chemicals, heavy metals, and algal toxins. An econometric model predicts the influence of water purification technology on water quality and human health. Reducing household water treated with water purification equipment can decrease up to 96% in predicted health benefits (Brown and Clasen 2012).

In summary, water contamination arises from a combination of anthropogenic and natural influences. Water quality can be directly impacted by a range of human activities, such as urbanisation, population increase, industrial output, climate change, and other variables, as well as religious activities. Improper disposal of solid debris, sand, and gravel contributes to the decline in water quality and quantity (Wang et al. 2024; Lin et al. 2022). The contaminants can disperse across several bodies of water as they are transported during various phases of the water cycle (Fig. 1.1).

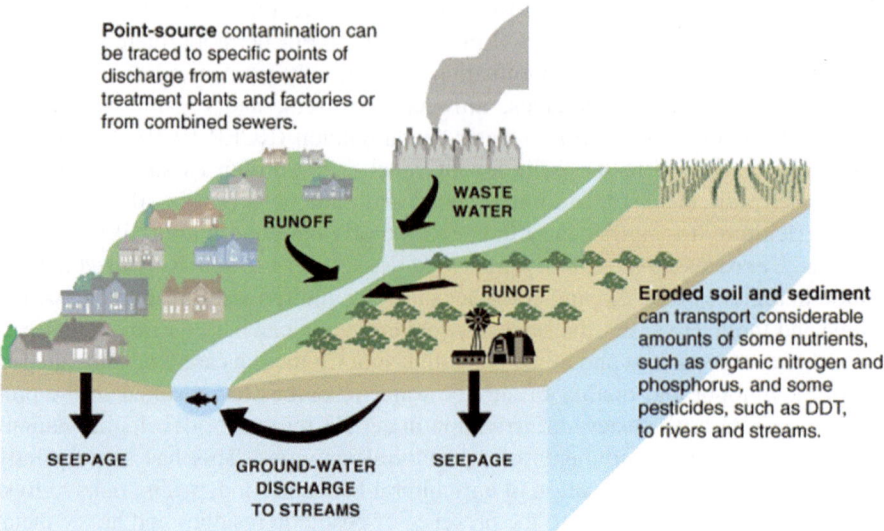

Fig. 1.1 Causes of water contaminants and their transfer into various water bodies along the water cycle. (The image is credited to USGS n.d. and is in the public domain)

1.3 Classification Based on Pollution Sources

The residence period, which refers to the average duration that a water molecule remains in a reservoir, plays a crucial role in pollution issues due to its impact on polluting potential. The residence time of water in rivers is relatively short, resulting in limited periods of pollution. Indeed, pollution in rivers has the potential to relocate to alternative bodies of water, such as the ocean, exacerbating further complications. Groundwater is frequently distinct by its slow movement and protracted stay, which can make groundwater contamination more problematic. Ultimately, the duration of pollution in a certain area can exceed the duration of water presence due to the possibility of pollutants being retained within the ecosystem or adhering to sediment for an extended period. Water can become polluted due to a range of human activities or the presence of naturally occurring mineral-rich geological formations. Potential sources of contamination include agricultural activities, industrial operations, landfills, animal operations, and both small and large-scale sewage treatment systems, among others. As water flows across the land or seeps into the earth, it chemically combines with and carries away substances deposited by various possible sources of contamination. The dangers and remediation methods for a pollutant are contingent upon its specific chemical composition.

Point-source pollution is caused by a specific and identifiable source. Animal factory farms, as an illustration, cultivate a substantial quantity of livestock, including cows, pigs, and chickens, in a densely populated environment. Combined sewer

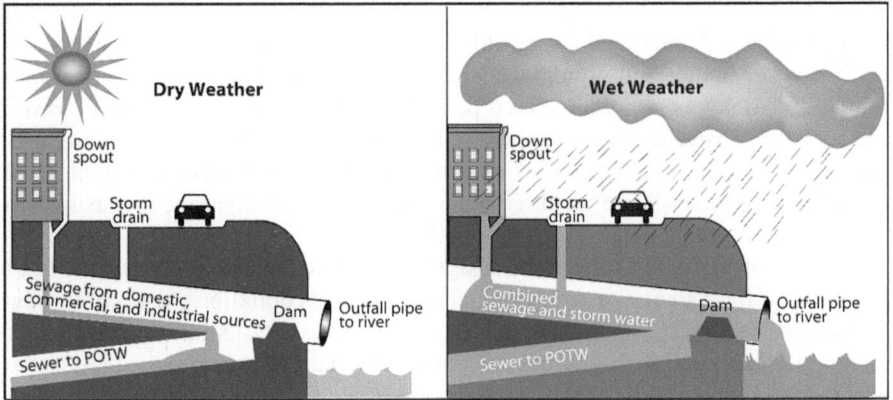

Fig. 1.2 Schematic of residential, commercial, and industrial storm water and sewage systems during dry periods, small storms, and heavy rainfall. Public treatment plants manage all sewage effectively during dry or minor storms, but excessive rain causes disintegration and pollutes neighbouring aquatic bodies. (The image is credited to the U.S. Environmental Protection Agency (EPA) and is in the public domain)

systems, which utilise a single network of underground pipes to gather both sewage and storm water runoff from roadways for wastewater treatment, can also serve as significant point sources of pollutants. During periods of intense rainfall, the volume of storm water runoff may surpass the capacity of the sewer system, resulting in a backup. This releases raw sewage directly into bodies of water without any treatment. Additional instances comprise conduits originating from industrial facilities, waste management locations, reservoirs, and incidents involving the release of hazardous substances. Figure 1.2 illustrates a combined sewer system where untreated sewage overflow after heavy rain is identified as a potential significant point cause of water contamination. Waste from households, businesses, and factories flows into the system via the down spout, while rainwater enters through the storm drain. The publicly owned treatment works (POTW) are in charge of managing all sewage during periods of aridity (as well as brief precipitation events). A dam serves as a barrier that keeps the unprocessed waste material, known as raw sewage, from entering a water body via the outfall pipe. In inclement weather conditions, specifically during heavy storms, the relief structure permits the release of a portion of the combined sewage and storm water without undergoing any treatment, directly into a nearby aquatic body.

Nonpoint-source contamination arises from numerous widely distributed sources. The collective presence of contaminants is detrimental, yet the individual constituents may not attain hazardous levels. Urban areas, abandoned mines, and agricultural fields are examples of non-point sources of contamination. Rainfall travels across the landscape and seeps into the soil, accumulating pollutants from the entire hydrological basin (including land areas and smaller streams that empty into a specific body of water). The contaminants could include fertilizers, herbicides, and pesticides from lawns and agricultural fields; road salt, oil, antifreeze, and animal feces from cities; and harmful materials and acid from abandoned min-ing operations (Ahmed and Ismail 2018). Rainfall then carries the pollutant into surface water bodies and groundwater. The main cause of water contamination in the United States is nonpoint-source pollution, which is more difficult to prevent and more expensive than point-source pollution because of its diluted nature, mul-tiple sources, and greater volumes of water (US-EPA n.d.). Chemical, biological, and physical processes are the three main ways in which impairment-causing sub-stances (water contaminants) are classified.

Transboundary water pollution is a significant issue involving the release of pol-lutants into water sources across national borders. This pollution can occur through factors like factories, agricultural runoff, and household tasks like laundering gar-ments or cleaning dishes. The contamination can harm aquatic ecosystems, cause health issues, and negatively impact economies dependent on fishing and tourism. To effectively address this issue, international collaboration is needed to design and enforce pollution prevention legislation and ensure adequate wastewater treatment infrastructure to purify contaminated water before discharge into the surrounding ecosystem.

1.4   Toxicology of Pollutants

The World Health Organisation (WHO) highlights potable water quality as a crucial determinant of human health. Inadequate water quality contributes to 80% of global diseases and 50% of child mortality worldwide. Over 50 diseases can be directly linked to poor drinking water quality. The potability of drinking water in underde-veloped nations is a concern, as water contamination remains the primary source of illness and death in developing nations. Research focuses primarily on the impact of water pollution on human health, specifically diarrhoea, skin illnesses, cancer, and child health (Warren-Vega et al. 2023). We have compiled the primary effects of water pollution on human health in Table 1.1. Impact of water pollution on diseases such as diarrhoea, skin illnesses, cancer, and child health.

Table 1.1 specifically examines the fields of environmental science, health, and medical literature. It primarily focuses on epidemiological studies that establish a connection between water quality, water pollution, and human diseases. Additionally, it summarizes the incidence and severity of water-related diseases. Simultaneously, it is crucial to prioritise publications from the United Nations and the World Health Organisation about studies on water and sanitation health. Table 1.1 aims to eluci-date the correlation between water pollution and human health, specifically focus-ing on the following aspects: the association between water pollution and diarrhoea, the connection between water pollution and skin diseases, the link between water pollution and cancer, and the impact of water pollution on child health.

Water pollution significantly impacts human health, causing various diseases. Enteroviruses, which are transmitted through aquatic environments, are the most easily caused by water pollution (Warren-Vega et al. 2023; Noor et al. 2023).

Table 1.1 Concise comprehensive research on the consequences of water contamination on human health

Country/region	Objective	Method	Significant results	Refs.
Water pollution and Diarrhoea				
Worldwide	Health benefits from water quality interventions require high adherence.	Quantitative microbial risk model	If more people had access to safe water, more than 500 DALYs might not happen for every 100,000 person-years. This is assuming that the water quality before treatment is fairly bad and that people follow the treatment instructions fully.	Brown and Clasen (2012)
Bangladesh	The link between water contamination and adverse health effects	The standard participatory rural appraisal (PRA) methods	People in the area are experiencing a host of health issues, including diarrhoea, as a result of river pollution.	Halder and Islam (2015)
Low-income or middle-income countries	Measures taken to enhance the purity of water	Meta-analysis	Diarrhoea rates can be decreased with actions aimed at enhancing water quality.	Clasen et al. (2015)
Delhi	Water contamination, including its causes, consequences, and control measures	A sample survey	Water contamination is linked to diarrhoea, according to the study.	Ahmed and Ismail (2018)
India	The effects of polluted water on people's Well-being	–	Ingesting parasitic worms found in water that does not undergo purification processes can lead to many ailments in humans.	Ansari and Akhmatov (2020)
Skin diseases caused by polluted water				
India	Human health consequences of water pollution	–	Local skin problems are caused by the usage of contaminated groundwater.	Kumar et al. (2017)
Worldwide	Signs and symptoms pertaining to the skin that occur after being in recreational water	Meta-analysis	Skin symptoms were shown to be associated with bacterial levels in seawater.	Yau et al. (2009)
Bangladesh	Effects of pollution in the workplace on people's Well-being	A sampling survey	Wastewater and water contamination due to growing industrial waste is directly associated with an increase in skin diseases.	Rabbani et al. (2010)

(continued)

Table 1.1 (continued)

Country/region	Objective	Method	Significant results	Refs.
Worldwide	Human health and water contamination	Meta-analysis	Diseases caused by bacteria, viruses, and parasites, such as gastrointestinal and skin infections, are spreading through contaminated water.	Haseena et al. (2017)
Bangladesh	Polluted river water and its effects on ecosystems and human health	A sampling survey	River contamination causes scabies in many persons.	Hanif et al. (2020)
Cleveland	Issues with industrial water contamination	–	Severe illnesses, such as stunted growth and skin cancer, may be caused by industrial water contamination.	Arif et al. (2020)
Cancer and the contamination of water				
Denmark	The risk of colon cancer and nitrate in drinking water	Cox proportional hazards models	Levels of nitrates are associated with their carcinogenic potential.	Schullehner et al. (2018)
Ganga	Water contamination poses a possible risk to the health of local residents.	–	Heavy metal contamination and pesticide residues in water are known to cause cancer.	Dwivedi et al. (2018)
India	Pollutants in groundwater and drinking water and the prevalence of cancer	Questionnaire method, ICPMS and AAS	Significant contaminants in groundwater include lead, uranium, fluoride, and nitrate, which may cause cancer.	Kaur et al. (2021)
China	The impact of water contamination on the rate of oesophageal cancer death after a lag period	Linear and nonlinear models	Water contamination over the past 8 years is the main cause of the current oesophageal cancer mortality rate.	Xu et al. (2019)
China	Preventing water contamination for the sake of people and the environment	Bottomeup inventory analysis hybrid life cycle impact assessment	Regional cancer incidence is influenced by water contamination.	Chen et al. (2019)

(continued)

Table 1.1 (continued)

Country/region	Objective	Method	Significant results	Refs.
Child health and water contamination				
Malaysia	Effects of water contamination on human health	–	Everywhere in the globe, people of all ages are losing their lives to water-related illnesses and accidents.	Afroz and Rahman (2017)
Worldwide	The impact of pollution on children's health	Review	A significant proportion of fatalities in children due to pollution are attributed to respiratory and gastrointestinal illnesses resulting from the contamination of air and water.	Landrigan et al. (2019)
China	Considerations for health risk assessment, water quality, sources, and characteristics	Principal component analysis	Water contamination disproportionately affects children.	Xiao et al. (2019)
Bulgarian	Kidney stones caused by nitrate contamination in drinking water	A comparative analysis	Goitre in children can be caused by nitrate poisoning in drinking water.	Vladeva et al. (2000)
India, Vietnam, and 33 African countries	The lingering consequences of water contamination on human capital: The nitrogen legacy	Panel regression analyses	Exposure to pollution during the formative years is linked to shorter stature in adulthood.	Zaveri et al. (2020)

Treatment for these diseases is crucial, as they can be prevented through drinking water interventions. Exposure to excessively contaminated water can lead to skin problems due to the infectious nature of high levels of bacteria and heavy metals in drinking water. Risks to one's health might arise from water contamination in the sources, treatments, and distribution systems (Ahmed and Ismail 2018; Chowdhary et al. 2020). Major carcinogens in water sources include trihalomethane, arsenic, nitrate, and chromium. Chlorine disinfection often introduces these carcinogens into the water. Complex factors, such as heavy metals, radionuclides, pesticides, herbicides, and chlorine byproducts, can pollute drinking water and cause cancer. In addition to being a major contributor to malnutrition, diarrhoea, and decreased immunity, water contamination also poses a substantial risk to children's health (Ahmed and Ismail 2018; Chowdhary et al. 2020).

1.5 Effects on Ecosystem

Water contamination has been a prominent global concern in recent years, garnering significant attention in the media. From unhygienic facilities in developing countries to the vast accumulation of plastic waste in the Great Pacific Garbage Patch to the pollution of drinking water sources, Water contamination is a significant global issue. However, what are the concrete manifestations of water pollution effects? Here is an in-depth examination of the impact of polluted water on humans, animals, plants, and the broader ecosystem (Envirotech n.d.).

1.5.1 Loss of Potable Water Sources

Given that our continued existence relies on it, it is only reasonable to elevate the quality and cleanliness of our drinking water sources above anything else. This is particularly accurate considering the fact that the global population is consistently increasing at a concerning pace, indicating that larger quantities of water will be needed in the future. Regrettably, the pollution of our lakes, reservoirs, and other water bodies has resulted in a diminishing supply of potable water for us to utilise.

1.5.2 Health Problems

In the event that drinking water supplies are insufficient to fulfil the demand, individuals will be compelled to use hazardous sources of drinking water. This is particularly evident in economically disadvantaged regions of the world, including Africa, Asia, and Latin America. According to the United Nations, almost one-third of the global population lacks access to this fundamental human right and essential resource. As a result, there are an unknown number of early fatalities in these areas due to watery illnesses like cholera, dysentery, and typhoid (Warren-Vega et al. 2023).

1.5.3 Eutrophication

Agricultural products, including fertilisers and pesticides, include high concentrations of ammonia and phosphate, which have the potential to enhance crop productivity. Nevertheless, these essential nutrients can enter streams, lakes, and other water channels via run-off, disrupting fragile aquatic ecosystems. This initiates a phenomenon known as eutrophication, in which specific organisms, such as algae, can rapidly multiply and absorb an excessive amount of oxygen and sunlight, thereby depriving other organisms of these vital nutrients.

1.5.4   Disruption of the Food Chain

Eutrophication can lead to the decline or extinction of specific species, which in turn can have a cascading impact on their predators, resulting in repercussions further up in the food chain. Simultaneously, the ingestion and assimilation of contaminated water by marine and aquatic creatures might result in harm to their interior organs. When larger animals devour them at a later stage, the food chain can be damaged and poisoned, ultimately impacting the human race adversely.

1.5.5   Farming

It is a well-established fact that agriculture consumes up to 70% of the Earth's water resources. In the event of water contamination, the limited availability of reserves for crop cultivation and livestock sustenance leads to decreased yields and lower crop quality. Furthermore, the use of contaminated water in the cultivation of plants and the rearing of animals renders them unsuitable for human consumption, indirectly exacerbating the detrimental effects on human health.

1.6   Global Laws on Water Pollution

Water pollution has been a crucial factor in the economic and social progress of early civilizations, with water being used for domestic consumption and navigation. The non-navigational uses of international watercourses after the Industrial Revolution influenced the development of international water law. The field of international watercourse law is gaining significance, but the main issue lies in the interpretation of provisions in relation to international legal rights and obligations. Positive treaty law may lead to uncertainty, while defining water pollution solely in terms of the harm it causes may only provide a clear understanding of its legal ramifications in light of international responsibility, particular circumstances, and fairness (Florio 1980).

The 2000 Ministerial Declaration of the Hague on Water Security identified several key challenges in achieving water security, including meeting basic needs, ensuring food security, preserving ecosystems, managing water resources, mitigating risks, recognising the value of water, and implementing effective water governance (Declaration 2000). UN Water has recently offered a definition of water security, which encompasses multiple elements related to the security of water. The discussion on legal security encompasses both legal and non-legal elements. Tadesse Kassa Woldestadik distinguishes between the former, which pertains to the legally enforceable rights of an individual or state, and the latter, which pertains to a reliable physical supply of water, whether it is legally allocated or obtained by capture

(Woldetsadik 2015). Furthermore, the issue of water security may be of significance to individuals, nations, and ecosystems. According to Wouters, Vinogradov, and Magsig, the fundamental concerns in water security may be summarised into three main topics: the presence of water, the ability to obtain water, and disputes related to water use (Wouters et al. 2009).

This subsection emphasizes the importance of a comprehensive comprehension of water security, focusing on the role and significance of international water law. It explores the complex relationship between humans and water, examining how international water law addresses the interaction between parties involved. The subsection also analyses the evolution of significant rules pertaining to water in global law and environmental law.

1.6.1 Maintaining Sovereignty: Water in Its Dual Role as Subject and Object

International law, originating from the Treaty of Westphalia in 1648, focuses on sovereignty as a legal framework for relations between independent states. Kennedy (1986) highlights the importance of jurisdictional concepts in addressing disputes between nations. The Harmon doctrine and the Lac Lanoux case illustrate conflicts between countries located upstream and downstream in terms of water resources (Kuokkanen 2017). These disputes can lead to potential clashes, as seen in the Harmon doctrine and the Lac Lanoux case, which successfully resolved disputes in a legally valid way. Despite the lack of practical solutions, international law continues to play a crucial role in ensuring the rights and interests of states.

In 1895, Mexican Minister Matias Romero criticized the diversion of water from the Rio Grande by farmers in Colorado and New Mexico, arguing it was diminishing the water supply for Mexican communities that relied on irrigation. The United States Secretary of State forwarded the matter to Judson Harmon, the Attorney General, for legal assessment. Harmon did not challenge Romero's claim regarding the reduction of water, but emphasized the issue of insufficient water for irrigation in both countries and posed the question of which country should prioritize over the other. Harmon argued that absolute sovereignty was the dominant concept in international law, stating that the core concept of international law is the complete authority of any nation within its territory in relation to all other nations. The Attorney General concluded that the United States was not subject to any liability or obligation under international law (Kuokkanen 2017).

The United States and Mexico faced water scarcity due to the Rio Grande's waters. Judson Harmon's "Harmon doctrine" outlined the concept of absolute sovereignty, where a state can control a river's portion without legal responsibility to downstream states. However, this doctrine also entails territorial integrity, allowing states to refuse harm from another state's territory. The Lac Lanoux case highlighted the effectiveness of international law in addressing such matters, as the doctrine's unequivocal nature was the main issue. Despite its historical significance, the Harmon doctrine remains a significant legal concept in the United States and Mexico. Another example is the arbitral tribunal involved in the Lac Lanoux case, which was a dispute between France and Spain (Kuokkanen 2017).

The theme of the above examples is that it is justifiable for a state to protect its water-related concerns by asserting its sovereignty. However, as the Lac Lanoux award and subsequent rulings show, sovereignty is not entirely unrestricted. Furthermore, general theories are neutral and do not inherently support or oppose any specific ideas. Depending on the particular circumstances and the relevant legal framework, these factors can result in either a favourable or unfavourable outcome in relation to water-related matters. Similarly, the connection between humanity and water is not a subject of concern for the broader field of global law (Kuokkanen 2017).

1.6.2   Protecting the Availability and Quality of Water Supplies: Water in Its Material Form

With the increase in water utilisation following World War II, a multitude of competing interests started to arise. Apprehension regarding the escalating conflicts and disputes among states began to intensify. General international law often lacked enough guidance, necessitating the establishment of substantive regulations to govern the use of water and prevent conflicts. Subsequently, during the 1960s and 1970s, water pollution emerged. Substantial rules were necessary to safeguard the waters for this specific objective as well. The regulations applied to water encompassed both its utilisation and protection. The establishment of the first international waterway administration occurred in 1804 to address navigation issues on the Rhine (Convention 1804). Furthermore, the process of internationalisation was expanded to include additional rivers (Kuokkanen 2017). The Statute on the Régime of Navigable Waterways of International Concern was approved in Barcelona in 1921 under the supervision of the League of Nations (Kuokkanen 2017). Initially, navigation was given priority over other concerns during the early days. However, over time, states began to acknowledge their interest in ensuring the non-navigational use of water as well. Prior to World War II, governments had already reached several bilateral and multilateral accords concerning non-navigational uses of boundary waters.

Given the various applications of water, conflicts often emerged when multiple users clashed, prompting the need for resolution strategies. According to Article 10 of the Convention on Non-Navigational Uses of International Watercourses (UN Convention 1997) unless there is an agreement or established practice stating otherwise, no particular use of water has a natural advantage over other uses. As per the article, the resolution of such a conflict should be based on the principles outlined in articles 5 (fair and reasonable utilisation and participation), article 6 (factors that are relevant to fair and reasonable utilisation), and article 7 (the duty to avoid causing significant harm), with particular consideration given to the requirements of essential human needs. Consequently, there has been a shift in focus from water usage for navigation and other purposes to prioritising essential human require-ments. With the rise of water pollution issues, states have acknowledged the impor-tance of safeguarding both the quantity and quality of water. While a few early border waters accords did address water preservation (Kuokkanen 2017), it was only during the 1960s and 1970s that several bilateral and multilateral conventions were enacted to safeguard international watercourses. Regulations were imple-mented to safeguard Lake Constance, the Mosel, the Rhine, and the Great Lakes (Kuokkanen 2017). These regulations specify precise targets for water quality or limitations on emissions. Alternatively, they create joint organisations that have the authority to determine specific regulations.

The theme of the above examples is that certain regulations were put in place to ensure the proper use of water for both navigation and non-navigation purposes, as well as to safeguard watercourses. Although both sets of regulations were centred around water, they had different objectives. The regulations for navigation primarily aimed to ensure an adequate supply of water, while the regulations for non-navigation focused on maintaining water quality. As challenges increased, it was acknowledged that there was a necessity to expand the range of regulations in order to address water security concerns (Woldetsadik 2015; Kuokkanen 2017).

1.6.3   Preserving Water for Future Generations: Water as Both an Object and a Subject

HA Smith's work on international water policy in the 1980s and 1990s emphasized the importance of sustainable water use and community ties (ROC 1929). This approach shifted the focus from quantity and quality factors to ensuring sustainable water use. Policymakers adopted a proactive approach to regulating freshwater eco-logical processes and adopted a pragmatic stance towards water-related concerns. Risk and crisis management strategies were implemented in water-related matters, incorporating environmental considerations into security matters. The Permanent Court of Justice also highlighted the shared interest in the River Oder case (ROC 1929), emphasizing the need for a shared interest among communities in water policy. Smith's work highlights the importance of fostering a sense of shared inter-est among communities, as seen in the case of the river Oder.

The UNECE Convention on the Protection and Use of Transboundary Watercourses and International Lakes is a commendable example of an environ-mental framework that aims to balance economic interests and environmental con-siderations (UNECE 1992; Kuokkanen 2017). This framework has been instrumental in integrating policy and scientific approaches to mitigate potential long-term detri-mental repercussions in water management. However, the conflict between exploi-tation and protection, as well as between specific and general objectives, such as immediate and long-term interests or human and aquatic environments, persists.

The new approach has introduced a range of problems, interests, and tensions that must be addressed within the same framework, highlighting the need for a more comprehensive approach to managing water resources. The UNECE Convention serves as a commendable example of how environmental considerations can be integrated into various industries, fostering sustainable development and addressing the challenges of balancing economic interests and environmental considerations.

The new method also adopted a more pragmatic perspective on water. Instead of solely aiming to mitigate water-related risks, it was acknowledged that certain severe occurrences, like floods and droughts, seem to be beyond human control and that climate change will exacerbate these extreme events. Consequently, states-initiated efforts to investigate strategies for enhancing readiness and managing the negative consequences of these occurrences. The link between man and water grew more ambiguous. Interestingly, water was no longer just something to be used or preserved but also a possible danger to humans (Wu et al. 2020).

1.7 Conclusions

The issue of water contamination requires the establishment of efficient rules and mechanisms to oversee and carry out efforts aimed at resolving the problem. This chapter examines the diverse factors that contribute to water contamination and classifies them according to the sources of pollution. The text provides a comprehensive analysis of the toxicological effects of water pollution, including its association with diarrhoea, skin ailments resulting from dirty water, the link between water contamination and cancer, and the impact on child health. Detailed analysis of the repercussions of contaminated water on human beings, animals, plants, and the wider ecosystem. Ultimately, we examine the worldwide regulations governing three overarching methods utilised by international law to address water-related matters: general international law, the regulatory approach, and the management approach. The chapter has aimed to illustrate the relevance of all these factors to water security.

References

Afroz R, Rahman A (2017) Health impact of river water pollution in Malaysia. Int J Adv Appl Sci 4(5):78–85. https://doi.org/10.21833/ijaas.2017.05.014
Ahmed S, Ismail S (2018) Water pollution and its sources, effects & management: a case study of Delhi. *Shahid Ahmed and Saba Ismail (2018) 'Water pollution and its sources, effects & management: a case study of Delhi'*. Int J Curr Adv Res 7(2):10436–10442
Ansari ZZ, Akhmatov SV (2020) Impacts of water pollution on human health: a case study of Delhi
Arif A, Malik MF, Liaqat S, Aslam A, Mumtaz K, Afzal A et al (2020) 3. Water pollution and industries. Pure Appl Biol (PAB) 9(4):2214–2224. https://doi.org/10.19045/bspab.2020.90237
Brown J, Clasen T (2012) High adherence is necessary to realize health gains from water quality interventions. PLoS One 7(5):e36735. https://doi.org/10.1371/journal.pone.0036735

Case Relating to the Territorial Jurisdiction of the International Commission of the River Order Permanent Court of Justice Judgement No 16 of 10 September 1929 (River Oder case) 27

Chen B, Wang M, Duan M, Ma X, Hong J, Xie F et al (2019) In search of key: protecting human health and the ecosystem from water pollution in China. J Clean Prod 228:101–111. https://doi.org/10.1016/j.jclepro.2019.04.228

Chowdhary P, Bharagava RN, Mishra S, Khan N (2020) Role of industries in water scarcity and its adverse effects on environment and human health. In: Environmental concerns and sustainable development: volume 1: air, water and energy resources, pp 235–256. https://doi.org/10.1007/978-981-13-5889-0_12

Clasen TF, Alexander KT, Sinclair D, Boisson S, Peletz R, Chang HH et al (2015) Interventions to improve water quality for preventing diarrhoea. Cochrane Database Syst Rev 10. https://doi.org/10.1002/14651858.CD004794.pub3

Convention on the Law of the Non-navigational Uses of International Watercourses (1997) UN Watercourse Convention

Convention Respecting the Navigation of Rhine between the Empire and France (1804)

Declaration HM (2000) Ministerial declaration of the Hague on water security in the 21st century. The Hague

Dwivedi S, Mishra S, Tripathi RD (2018) Ganga water pollution: a potential health threat to inhabitants of ganga basin. Environ Int 117:327–338. https://doi.org/10.1016/j.envint.2018.05.015

Envirotech (n.d.). https://www.envirotech-online.com/news/water-wastewater/9/breaking-news/6-effects-of-water-pollution/57707

Florio F (1980) Water pollution and related principles of international law. Canadian Yearbook Int Law 17:134–158. https://doi.org/10.1017/S0069005800001466

Halder JN, Islam MN (2015) Water pollution and its impact on the human health. J Environ Hum 2(1):36–46. https://doi.org/10.15764/eh.2015.01005

Hanif MA, Miah R, Islam MA, Marzia S (2020) Impact of Kapotaksha river water pollution on human health and environment. Progress Agric 31(1):1–9. https://doi.org/10.3329/pa.v31i1.48300

Haseena M, Malik MF, Javed A, Arshad S, Asif N, Zulfiqar S, Hanif J (2017) Water pollution and human health. Environ Risk Assess Remediat 1(3). https://doi.org/10.4066/2529-8046.100020

Kaur G, Kumar R, Mittal S, Sahoo PK, Vaid U (2021) Ground/drinking water contaminants and cancer incidence: a case study of rural areas of south West Punjab, India. Hum Ecol Risk Assess Int J 27(1):205–226. https://doi.org/10.1080/10807039.2019.1705145

Kennedy D (1986) Primitive legal scholarship. Harv Int LJ 27:1

Kumar S, Meena HM, Verma K (2017) Water pollution in India: its impact on the human health: causes and remedies. Int J Appl Environ Sci 12(2):275–279

Kuokkanen T (2017) Water security and international law. Potchefstroom Electron Law J 20(1). https://doi.org/10.17159/1727-3781/2017/v20i0a1652

Landrigan PJ, Fuller R, Fisher S, Suk WA, Sly P, Chiles TC, Bose-O'Reilly S (2019) Pollution and children's health. Sci Total Environ 650:2389–2394. https://doi.org/10.1016/j.scitotenv.2018.09.375

Lin L, Yang H, Xu X (2022) Effects of water pollution on human health and disease heterogeneity: a review. Front Environ Sci 10:880246. https://doi.org/10.3389/fenvs.2022.880246

Lipponen (1992) The UNECE water convention. The convention has been strengthened by the Protocol on Water and Health to the 1992 Convention on the Protection and Use of Transboundary Watercourses and International Lakes (1885) (Water and Health Protocol)

Lu Y, Song S, Wang R, Liu Z, Meng J, Sweetman AJ et al (2015) Impacts of soil and water pollution on food safety and health risks in China. Environ Int 77:5–15. https://doi.org/10.1016/j.envint.2014.12.010

Nationen V (2003) Water for people, water for life: the United Nations world water development report; a joint report by the twenty-three UN agencies concerned with freshwater. UNESCO Public Library Manifesto

Noor R, Maqsood A, Baig A, Pande CB, Zahra SM, Saad A et al (2023) A comprehensive review on water pollution, South Asia region: Pakistan. Urban Clim 48:101413

Parris K (2014) Impact of agriculture on water pollution in OECD countries: recent trends and future prospects. Water Qual Manag:33–52. https://doi.org/10.1080/07900627.2010.531898

Rabbani MG, Chowdhury M, Khan NA (2010) Impacts of industrial pollution on human health: empirical evidences from an industrial hotspot (Kaliakoir) in Bangladesh. Asian J Water Environ Pollu 7(1):27–33

Richard C (2021) The United Nation World Water Development Report 2021: valuing water; executive summary. UNESCO World Water Assessment Programme

Royal Commission on Environmental Pollution (1979) Seventh Report: Agriculture and Pollution

Schullehner J, Hansen B, Thygesen M, Pedersen CB, Sigsgaard T (2018) Nitrate in drinking water and colorectal cancer risk: a nationwide population-based cohort study. Int J Cancer 143(1):73–79. https://doi.org/10.1002/ijc.31306

US-EPA (n.d.). https://commons.wikimedia.org/wiki/File:CSO_diagram_US_EPA.svg

USGS (n.d.). https://pubs.usgs.gov/circ/circ1225/html/sources.html

Vladeva S, Gatseva P, Gopina G (2000) Comparative analysis of results from studies of goitre in children from Bulgarian villages with nitrate pollution of drinking water in 1995 and 1998. Cent Eur J Public Health 8(3):179–181

Wang M, Bodirsky BL, Rijneveld R, Beier F, Bak MP, Batool M et al (2024) A triple increase in global river basins with water scarcity due to future pollution. Nat Commun 15(1):880

Warren-Vega WM, Campos-Rodríguez A, Zárate-Guzmán AI, Romero-Cano LA (2023) A current review of water pollutants in American continent: trends and perspectives in detection, health risks, and treatment technologies. Int J Environ Res Public Health 20(5):4499

Woldetsadik TK (2015) Remodeling sovereignty: overtures of a new water security paradigm in the nile basin legal discourse. Sovereignty and the Development of International Water Law, 8. Available at SSRN: https://ssrn.com/abstract=2764619

Wouters P, Vinogradov S, Magsig BO (2009) Water security, hydrosolidarity, and international law: a river runs through It…. In: Hydrosolidarity, and international law: a river runs through it, p 97–134. Available at SSRN: https://ssrn.com/abstract=2365328

Wu H, Gai Z, Guo Y, Li Y, Hao Y, Lu ZN (2020) Does environmental pollution inhibit urbanization in China? A new perspective through residents' medical and health costs. Environ Res 182:109128. https://doi.org/10.1016/j.envres.2020.109128

Xiao J, Wang L, Deng L, Jin Z (2019) Characteristics, sources, water quality and health risk assessment of trace elements in river water and well water in the Chinese loess plateau. Sci Total Environ 650:2004–2012. https://doi.org/10.1016/j.scitotenv.2018.09.322

Xu C, Xing D, Wang J, Xiao G (2019) The lag effect of water pollution on the mortality rate for esophageal cancer in a rapidly industrialized region in China. Environ Sci Pollut Res 26:32852–32858. https://doi.org/10.1007/s11356-019-06408-z

Xu X, Yang H, Li C (2022) Theoretical model and actual characteristics of air pollution affecting health cost: a review. Int J Environ Res Public Health 19(6):3532. https://doi.org/10.3390/ijerph19063532

Yau V, Wade TJ, de Wilde CK, Colford JM (2009) Skin-related symptoms following exposure to recreational water: a systematic review and meta-analysis. Water Qual Expo Health 1:79–103. https://doi.org/10.1007/s12403-009-0012-9

Zaveri ED, Russ JD, Desbureaux SG, Damania R, Rodella AS, Ribeiro Paiva De Souza G (2020) The nitrogen legacy: the long-term effects of water pollution on human capital. World Bank Policy Research Working Paper (9143). Available at SSRN. https://ssrn.com/abstract=3533562

Chapter 2
Assessment of Different Types of Water Pollution Redressal

Abstract Water pollution continues to be a worldwide issue that affects both human and environmental well-being. There is a need to consistently tackle the rising problems of pollution to ensure a pristine and secure subterranean environment for residents. Therefore, certain categories of water pollutants that cause significant water pollution and toxicity necessitate the use of biological processes instead of physico-chemical methods to treat the contaminated water, which contains a variety of pollutants. This is crucial in order to protect the environment and human health. The Water Quality Index (WQI) is a widely used instrument for assessing and characterizing water quality. Physical, chemical, and biological elements combine to determine the value, which ranges from 0 to 100. This measurement involves four distinct processes. The present chapter focuses on assessing water contamination caused by a variety of pollutants.

Keywords Assessment · Water pollution · Water quality index · Redressal

2.1 Introduction

Water is essential and crucial for the survival of human beings. The rapid progress of human society has resulted in water contamination, with a wide range of pollutants present in varying quantities. The presence of pollutants in water can lead to negative economic impacts, health issues in humans, and ecological destruction. Therefore, there is an urgent need for thorough discussion on the analysis and mitigation techniques for water contamination.

There has been a tremendous uptick in the amount of focus on water quality assessment and pollution monitoring. There is a strong correlation between water contamination and both public health problems and the loss of natural habitats. Domestic, industrial, and agricultural operations are only a few of the many sources of the many hazardous components that make up wastewater. There has been a lot of focus on the decline in water quality due to rising levels of contaminants

N. Saxena et al., *Water Pollution and Remediation*, SpringerBriefs in Water Science and Technology, https://doi.org/10.1007/978-3-031-76301-4_2

such pesticides, heavy metals, and infectious and noncommunicable illnesses. Public health and habitat preservation can be spared via the prompt detection and assessment of these dangers. Treatment techniques, design, functionality, and water or waste reuse are all affected by the concentration of wastewater components and signs of infectious illnesses. In order to conduct remedial methods or monitor public health, it is necessary to monitor the concentrations of contaminants over time (Srivastava et al. 2022).

The industrial effluent is always emitted into the atmosphere, which poses serious health hazards to the environment and people (Yadav et al. 2017; Zainith et al. 2016). As a result of vast manufacturing discharge of industrial wastewater (IWW) without treatment, water bodies are threatening the sustainability and safety of the environment through pollution of natural resources (Gricic et al. 2009). Traditional treatment processes such as adsorption, precipitation, ion exchange, coagulation, biosorption, electro-dialysis (ED) etc., have been in practice in management of waste generated through industrial activities since ancient time (Narmadha and Kavitha, 2012). These are more costly and require more energy; therefore, they are less suitable and efficient. For increased effectiveness over the conventional procedures, new and effective approaches are being designed. These technologies offer superb analytical capability for characterizing wastewater. Due to the low cost, the ability to protect the environment, and the more effective disposal of hazardous waste, such renewable technologies as bioremediation and phytoextraction are increasingly being used to treat Industrial Waste Water (IWW) (Rajasulochana and Preethy 2016; Kanagare et al. 2016; Kumar and Gopal 2015).

Various methods such as chemical reaction (Yin et al. 2003), photo degradation (Bagade et al. 2023), and evaporation by heat (Smetana et al. 2005; Kuhn et al. 2002) are available to remove water. However, most of these methods employ dangerous reagents, need special apparatus, are energy-intensive, often expensive and entail labour intensive cleaning steps. A vast study has also been done on conventional contaminants, including heavy metals and nutrients. However, at the moment, the development of effective methods for analysis of these pollutants, especially for testing situations on-site and in natural conditions, presents a major challenge. Emerging pollutants include endocrine-disrupting chemicals, which are hazardous substances that affect both humans and ecosystems, even at low concentrations. Therefore, there is a desperate need to carry out more investigations and create new techniques to detect very small particles of pollutants in water. There is a challenge when it comes to the efficient analysis of pollutants in soil because of the pre-treatment methods that are normally prudent. Water is a crucial resource for the sustenance of human life and civilization, which has been threatened by water pollution, making it imperative for the development of proper means of water purification. Some questions remain regarding the economic characteristics of innovative ways to solve challenges and what is going to happen to all the contaminants once they are addressed together with a long list of factors that may influence the situation. In the present chapter authors have given general ideas assessment of the water quality index and about different types of water pollution and their assessment and remediation techniques.

2.2 Assessment of Water Quality

Water is a critical natural resource for human existence, with significant social and economic importance. Approximately 1.1 billion individuals lack access to potable water, and two-thirds of all countries are projected to face water scarcity by 2025 (Chidiac et al. 2023). Human activities, climate conditions, and hydrology contribute to the buildup of pollutants in surface water, degrading the water supply. Ensuring adequate quantity and satisfactory water quality is crucial for survival. Maintaining acceptable water quality, however, is a challenge in water resource management. The Water Quality Index (WQI) is a highly effective method for evaluating water quality, as it condenses a large amount of information into a single value ranging from 0 to 100 (Chidiac et al. 2023) as discussed in Table 2.1 (Ejaz et al. 2024). Assessing the quality of bodies of water entails examining physical, chemical, and biological attributes associated with human activities or natural events.

2.2.1 Water Quality Index (WQI)

The latest WQI model established in the literature was conducted in 2017. This index aimed to mitigate the ambiguity inherent in existing water quality indexes. The West Java Water Quality Index (WJWQI) was implemented in the Java Sea in Indonesia, utilising 13 essential factors to assess water quality (Sutadian et al. 2016). These variables include temperature, suspended particles, COD, DO, nitrite, total phosphate, detergent, phenol, chloride, Zn, Pb, mercury (Hg), and faecal coliforms. Through the implementation of two screening procedures utilising statistical assessment, it was discovered that there are only 9 parameters (variables) that exhibit redundancy. These parameters are temperature, suspended particles, COD, DO, nitrite, total phosphate, detergent, phenol, and chloride. Sub-indices were derived for the nine variables and weights were assigned based on expert judgements, utilising the identical multiplicative aggregation method as the NSFWQI. The WJWQI proposed a classification system consisting of 5 quality classifications, which span from low (5–25) to high (90–100) (Uddin et al. 2021).

Table 2.1 Classification of CCME-WQI and its accompanying water quality condition

Water quality class	Index range	Water status
Excellent	95–100	Very good quality water
Good	80–94	Good quality water
Fair	65–79	Acceptable quality water
Marginal	45–64	Poor quality water
Poor	0–44	Very poor-quality water

Ejaz et al. (2024)

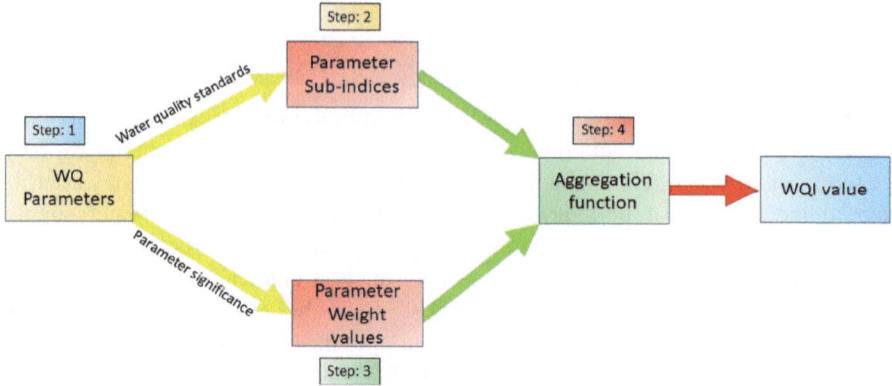

Fig. 2.1 The WQI model's broad framework is depicted in an image under CC licensing. (Uddin et al. 2021)

2.2.2 Various Steps in WQI Model Structure

The overall framework of WQI models is depicted in Fig. 2.1 and demonstrates that the majority of WQIs consist of four primary stages (Abbasi and Abbasi 2012; Abrahão et al. 2007; Lumb et al. 2011; Sutadian et al. 2016), chiefly:

1. The water quality parameters are selected for the evaluation: one or more parameters to be included.
2. The parameter sub-indices are generated: parameter concentrations are converted into unitless values.
3. Allocation of the parameter weight values: Weights are allocated to criteria based on their importance to the evaluation.
4. The individual parameter sub-indices are aggregated using the weightings to get a unified overall index. A grading system is commonly employed to categorise and classify water quality based on the aggregate index value.

2.3 Different Types of Water Pollution

2.3.1 Ground Water Pollution

Effective water resource management is a vital issue for the majority of governments worldwide. Only 3% of the Earth's total water is considered a potable source, with groundwater accounting for roughly 30% of this. Groundwater is critical for people's health, ecosystems, the energy sector, and other industries that rely on water. The exhaustion of groundwater and poor quality in many locations have resulted from the growing demand for residential, industrial, and agricultural use. From a sustainable development perspective, water contamination in any region of the world has environmental, economic, and social repercussions. Therefore, it is imperative to give meticulous consideration to the conservation of aquatic reserves (Kalhor et al. 2019).

Groundwater is a vital reserve of potable water for both humans and animals worldwide. Additionally, it serves as a crucial water source for drinking, agricultural, and industrial purposes. The hydrological cycle intricately links water resources, which rely on rainfall and recharging technologies. The physical, chemical, and biological properties of groundwater determine its quality. The quality of groundwater primarily determines its suitability for various purposes. Water quality evaluation is a crucial technique for disseminating information about the characteristics of potable water to the general population. It serves as a gauge for the water's purity. The objective of assessing the state of the water is to transform complex input data about the state of the water into concise details that are crucial for the general audience. Several separate studies conducted an examination of the state of groundwater by calculating the water quality index to support the understanding of the changes in the quality of groundwater (Mohan et al. 2019).

This study adopts an anthropocentric perspective by focusing on the sources of human-caused pollution, the effects of contamination on people, and the relevant oversight strategies. Anthropogenic activities primarily cause groundwater contamination. Groundwater is particularly susceptible in regions with high population density and intensive human land use. Practically Any action, whether intentional or accidental, that releases chemicals or waste into the environment has the potential to contaminate groundwater. Contaminated groundwater is a challenging and costly endeavour (Abel and Tosin 2019). The rapid growth of industry, urbanisation, and agricultural output has resulted in water scarcity in numerous regions across the globe. The water resources of the reservoir remain relatively stable as long as the need for water persists. India's available water reserves are 1123 BCM for surface water and 433 BCM for groundwater, according to the available estimates. Sivakumar and colleagues conducted the study in 2015. In order to tackle pollution prevention or remediation, it is crucial to comprehend the interconnection between groundwater and surface water. Recognizing the interconnection between surface water and groundwater is critical for a comprehensive understanding and effective management of these water resources. If a water supply well is located in close proximity to a source of contamination, there is a possibility that the well may become polluted (Fig. 2.2). In the event that there is a stream or river in close proximity, the groundwater contamination may extend to pollute that water body as well. A pollutant possesses the ability to pollute groundwater, considering its biological, chemical, and physical properties. Remediating contaminated groundwater is a challenging and costly endeavour (Abel and Tosin 2019).

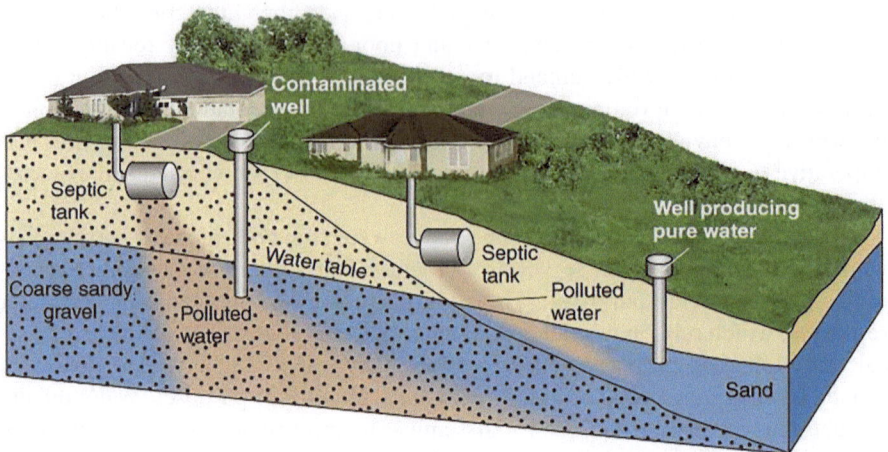

Fig. 2.2 The graphics depict the process of groundwater contamination. (Abel and Tosin 2019), and are made available under a Creative Commons license

2.3.2   Chemical Pollution

Chemical pollution is a worldwide environmental issue with extensive impacts on ecosystems, human well-being, and the entire planet. It denotes the emission of noxious substances into the surroundings, resulting in detrimental impacts on living creatures and ecological systems. This article will examine the origins and consequences of chemical pollution on the environment, as well as investigate possible remedies to reduce its harmful impacts.

Industrial processes are a major contributor to chemical contamination. Factories and manufacturing plants emit a wide range of chemicals and pollutants into the atmosphere, bodies of water, and land as a result of their production activities. The substances in question may consist of heavy metals, solvents, and poisonous gases. Heavy metals, such as lead, which is utilised in battery production and has the potential to pollute both the air and water; mercury, commonly present in fluorescent light bulbs; and cadmium, employed in electroplating processes. Solvents are utilised in the manufacturing processes in the fields of rubber, leather, adhesives, coatings, inks, and paints. Hydrogen sulphide, ammonia (refrigeration), and chlorine (water treatment) are examples of toxic gases. Industrial and agricultural activities entail the utilisation of several chemicals that have the potential to be washed away into water bodies, leading to water pollution.

Metals and solvents released during industrial activities have the potential to contaminate water bodies. These substances are toxic to several types of marine's organisms and can hinder their growth, impair their ability to reproduce, or even cause mortality.

- Pesticides are employed in agriculture to manage the growth of weeds, the presence of insects, and the spread of fungi. The discharge of these chemicals can result in water pollution and the contamination of aquatic organisms. Consequently, avian, human, and other animal species may experience poisoning if they consume contaminated fish.
- Petroleum, a type of toxic substance, commonly pollutes water sources due to oil spills caused by vessel breaks down. Disasters caused by oil leaks typically have a limited impact to animals in a specific area, but they can extend over long distances. The presence of oil can result in the mortality of numerous fish and adhere to the plumage of seabirds, impairing their flight capabilities.

Chemical processes such as flocculation, oxidation, precipitation, and algaecide can eliminate suspended solids (SS) and algal blooms from polluted water. Chemical procedures offer a rapid solution for treating contaminated river water (Peilin et al. 2019). However, these methods are only temporary and can generate additional waste, potentially leading to other risks. So, when flocculation or precipitation is done, chemicals that are safe for the environment should be used first to get rid of the suspended particles and algae. In their 2012 study, (Wang et al. 2012) found that poly aluminium chloride can serve as an environmentally friendly flocculation agent, efficiently separating algae from water. Hybrid remediation strategies frequently observe the adsorptive elimination of impurities from river water sources by minerals or material surfaces.

Literature has documented various ecologically based biological methods for treatment, such as biological remediation by microbes, biofilms, membrane bioreactor technology, ecological ponds, ecological floating beds, artificial wetlands, plant purification treatment, and contact oxidation (Bai et al. 2020). Polluted river water is characterised by its foul smell, cloudiness, reduced water clarity, elevated levels of chemical oxygen demand (COD), biological oxygen demand (BOD), and the presence of both organic and inorganic pollutants.

2.3.3 Microbial Pollution

Water is essential for sustaining life. Throughout human history, the provision of water has been a fundamental aspect of society, serving a multitude of purposes such as drinking, agriculture, industry, and domestic use. The lack of access to potable water is a significant contributing factor to the prevalent health issues in poor nations (Some et al. 2021). Estimates suggest that contaminated drinking water contributes to approximately 80% of all diseases and more than one-third of deaths in underdeveloped nations. The World Health Organisation (WHO) reported that around 600 million instances of diarrhoea and dysentery, along with 46,000 newborn deaths, occur annually due to contaminated water and inadequate sanitation (Singh et al. 2001). Surface water can be hazardous when it comes into contact with human waste or sewage from the surrounding area, as it can carry disease-causing germs. Microbially contaminated water poses significant risks for activities such as drinking, swimming, bathing, and other related purposes. In addition, the presence of sewage in water disrupts the oxygen equilibrium, posing a significant threat to aquatic plants and animals. We have monitored the water's integrity to determine whether a water supply is suitable for a specific use. The assessment of water appropriateness is based on the allowable concentrations of specific water quality characteristics, as established by recommendations, criteria, or the highest allowable level. Faecal microorganisms that cause diarrhoea in water indicate the water's history of contamination by faecal matter from endothermic organisms. This is why measuring the amount of feces and bacteria is important for both basic and applied research in the area of microbe ecology in water, as well as the development of parameter-driven technologies for checking the quality of drinking water. The presence of faecal bacteria like Klebsiella, streptococci, and Clostridium perfringens signifies the presence of faecal matter in the water. In addition to that, Salmonella typhi, Shigella spp., Salmonella typhimurium, Proteus spp., and Salmonella enteritidis are the most prevalent bacterial species found in sewage samples (Harwood et al. 2010). Several physicochemical parameters, including chemical oxygen demand (COD), total suspended solids (TSS), pH, turbidity, alkalinity, salinity, pH, and concentrations of heavy metals, influence the water quality (Some et al. 2021).

2.3.4   Nutrient Pollution

Excessive introduction of nutrients into a body of water leads to nutrient pollution, a type of water pollution. An excess of nutrients, typically nitrogen or phosphorus, primarily causes eutrophication of surface waters, such as lakes, rivers, and coastal waters, stimulating algae growth (Walters 2016). The primary sources of nutrient contamination are water discharge from agricultural emissions, runoff from feedlots and septic tanks, and burning processes on farms and pastures. Sewage, which contains excessive levels of nutrients, is a significant cause of cultural eutrophication. Discharging untreated human waste into a significant body of water is known as sewage dumping, and it continues to happen globally. Excessive reactive nitrogen molecules in the environment link to numerous significant environmental issues. The mentioned environmental issues encompass eutrophication of surface waters, detrimental algal blooms, hypoxia, acid rain, nitrogen saturation in forests, and climate change.

Due to two agricultural heydays between the 1910s and 1940s, driven by increased demand for food, agricultural output has become increasingly dependent on fertilizer usage (Seo et al. 2004). Fertiliser is a material, either natural or artificially altered, that enhances the fertility of soil. These fertilisers possess elevated levels of phosphate and nitrogen, leading to an excessive influx of dietary minerals into the soil. Phosphorus, nitrogen, and potassium are the three basic elements in commercial fertilisers, sometimes referred to as the "Big 3." Each of these essential minerals plays a critical role in plants' dietary requirements. If developing plants fail to fully utilize phosphorus and nitrogen, they may end up abandoning them in fields, which can lead to adverse effects on air and water purity (Rauh 2021). Such nutrients have the potential to eventually enter aquatic habitats and contribute to eutrophication escalation. When farmers apply fertilizer, no matter how natural or synthetic, a portion of it will wash away so that water can flow off and accumulate downhill, causing cultural enrichment.

2.3.5   Oxygen Depletion Pollution

Another outcome of nutrient water contamination is the reduction of oxygen levels. In conditions of insufficient oxygen, organism's dependent on oxygen for respiration will perish, while anaerobic species will thrive. Due to the production of ammonia and other noxious substances by numerous anaerobic organisms, water in such conditions becomes even more hazardous for both aquatic species and humans who depend on water sources for drinking.

The main factor leading to oxygen depletion in a water body is excessive growth of algae and phytoplankton due to elevated levels of phosphorus and nitrogen. The amounts of dissolved oxygen are influenced by factors such as temperature, lake depth, efficiency and fertility, and water movement. Fluctuations in typical nutrient cycles in almost any water-based ecosystem can cause imbalances that result in a decrease in oxygen levels and the death of fish (Welker et al. 2013).

2.3.6   Surface Water Pollution

The term "surface water" refers to the water that is found in our oceans, lakes, rivers, and other bodies of water that are portrayed as blue on maps. Surface water accounts for roughly 70 percent of the total surface area of the Earth (Singh et al. 2022). With the exception of the ocean, more than 60% of the water that is delivered to households in the United States originates from freshwater sources. However, a large percentage of that water is in danger of becoming contaminated. Approximately 50% of our rivers and streams, in addition to more than 33% of our lakes, have been determined to be polluted and unfit for activities such as swimming, fishing, and drinking, according to the most recent assessments that have been carried out by the Environmental Protection Agency of the United States of America. The most common type of contamination found in these freshwater sources is nutrient pollution, which includes both nitrates and phosphates. Despite the fact that plants and animals require these minerals for growth, they have become a substantial cause of pollution as a result of the discharge of fertilisers and the waste that is produced by farms. In addition, a considerable quantity of contaminants are released into the environment via municipal and industrial waste discharges. In addition to this, there is the garbage that is discharged into water bodies without any discrimination, which is a problem that is caused by both industry and individuals.

2.4 Conclusion

Water contamination is a consequence of human society's rapid advancement, which has negative effects on the economy, public health, and the environment. Emerging contaminants present substantial hazards to both humans and ecosystems. Evaluating water quality using the Water Quality Index (WQI) is crucial for determining the extent of water pollution. The WQI is a straightforward technique that quantifies water quality by considering a specific parameter and providing a single result. Groundwater contamination, microbiological pollution, nutrient pollution, surface water pollution, and chemical pollution have a substantial impact on a considerable amount of Earth's water resources. This is a serious concern due to the increasing demand for residential, industrial, and agricultural uses. We assess the quality of water based on its physical, chemical, and biological characteristics.

References

Abbasi T, Abbasi SA (2012) Water quality indices. Elsevier

Abel OT, Tosin JK (2019) Groundwater pollution and remediation. J Water Resource Prot 11:1–19

Abrahão R, Carvalho M, Da Silva Jr WR, Machado T, Gadelha C, Hernandez M (2007) Use of index analysis to evaluate the water quality of a stream receiving industrial effluents. Water SA 33(4)

Bagade AV, Pund SN, Nagwade PA, Kumar B, Deshmukh SU, Kanagare AB (2023) Ni-doped Mg-Zn nano-ferrites: fabrication, characterization, and visible-light-driven photocatalytic degradation of model textile dyes. Catal Commun 181:106719

Bai XY, Zhu XF, Jiang HB, Wang ZQ, He CG, Sheng LX, Zhuang J (2020) Purification effect of sequential constructed wetland for the polluted water in urban river. Water 12:1054

Chidiac S, El Najjar P, Ouaini N, El Rayess Y, El Azzi D (2023) A comprehensive review of water quality indices (WQIs): history, models, attempts and perspectives. Rev Environ Sci Biotechnol 22(2):349–395

Ejaz U, Khan SM, Jehangir S, Ahmad Z, Abdullah A, Iqbal M et al (2024) Monitoring the industrial waste polluted stream-integrated analytics and machine learning for water quality index assessment. J Clean Prod 450:141877

Grčić I, Vujević D, Šepčić J, Koprivanac N (2009) Minimization of organic content in simulated industrial wastewater by Fenton type processes: a case study. J Hazard Mater 170(2–3):954–961

Harwood VJ, Ryu H, Domingo JS (2010) Microbial source tracking. In: The Fecal Bacteria, pp 189–216

Kalhor K, Ghasemizadeh R, Rajic L, Alshawabkeh A (2019) Assessment of groundwater quality and remediation in karst aquifers: a review. Groundw Sustain Dev 8:104–121

Kanagare AB, Singh KK, Kumar GK, Kumar VSSM (2016) Synthesis of potassium nickel hexacyanoferrate encapsulated polymeric beads for extraction of cesium. Synthesis 5(1):14–20

Kanagare AB, Ajish JK, Singh KK, Bairwa KK, Shinde VS, Kumar M (2017) Asian J Mater Chem

Kuhn LT, Bojesen A, Timmermann L, Nielsen MM, Mørup S (2002) Structural and magnetic properties of core–shell iron–iron oxide nanoparticles. J Phys Condens Matter 14(49):13551–13567

Kumar BL, Gopal DS (2015) Effective role of indigenous microorganisms for sustainable environment. 3 Biotech 5:867–876

Lumb A, Sharma TC, Bibeault JF (2011) A review of genesis and evolution of water quality index (WQI) and some future directions. Water Qual Expo Health 3:11–24

Mohan S, Muralimohan N, Vidhya K, Sivakumar CT (2019) Assessment of ground water pollution and remediation–a review. Int Res J Multidiscip Technovation 1(6):650–657

Narmadha D, Kavitha VMS (2012) Treatment of domestic waste water using natural flocculants

Peilin G, Meng C, Lichao Z, Yuejun S, Minghao M, Lingyun W (2019) Study on water ecological restoration technology of river. IOP Conf Ser Earth Environ Sci 371(3):032025

Rajasulochana P, Preethy V (2016) Comparison on efficiency of various techniques in treatment of waste and sewage water–a comprehensive review. Resou Effic Technol 2(4):175–184

Rauh E (2021) Exploring intensive agriculture and organic fertilizer management in the US: implications for nutrient pollution prevention. Arizona State University

Seo S, Aramaki T, Hwang Y, Hanaki K (2004) Environmental impact of solid waste treatment methods in Korea. J Environ Eng 130(1):81–89

Shivajirao PA (2012) Treatment of distillery wastewater using membrane technologies. Int J Adv Eng Res Stud 1(3):275–283

Singh RK, Iqbal SA, Seth PC (2001) Bacteriological pollution in a stretch of river Narmada at Hoshangabad, Madhya Pradesh. Pollut Res 20(2):211–213

Singh R, Andaluri G, Pandey VC (2022) Cities' water pollution – challenges and controls. In: Algae and aquatic macrophytes in cities. Elsevier, pp 3–22

Sivakumar CT, Tamilchelvan P, Mohan S, Silambarasan D (2015) Monthly variations in ground water quality in Mallasamudram village and adjacent areas, Namakkal District, Tamil Nadu, India (post monsoon season). Int J Appl Eng Res 10(13):11733–11746

Smetana AB, Klabunde KJ, Sorensen CM (2005) Synthesis of spherical silver nanoparticles by digestive ripening, stabilization with various agents, and their 3-D and 2-D superlattice formation. J Colloid Interface Sci 284(2):521–526

Some S, Mondal R, Mitra D, Jain D, Verma D, Das S (2021) Microbial pollution of water with special reference to coliform bacteria and their nexus with environment. Energy Nexus 1:100008

Srivastava P, Mittal Y, Gupta S, Abbassi R, Garaniya V (2022) Recent progress in biosensors for wastewater monitoring and surveillance. In: Artificial intelligence and data science in environmental sensing, pp 245–267

Sutadian AD, Muttil N, Yilmaz AG, Perera BJC (2016) Development of river water quality indices – a review. Environ Monit Assess 188:1–29

Uddin MG, Nash S, Olbert AI (2021) A review of water quality index models and their use for assessing surface water quality. Ecol Indic 122:107218

Walters A (ed) (2016) Nutrient pollution from agricultural production: overview, management and a study of Chesapeake Bay. Nova Science Publishers, Hauppauge

Wang J, Liu XD, Lu J (2012) Urban river pollution control and remediation. Procedia Environ Sci 13:1856–1862

Welker AF, Moreira DC, Campos ÉG, Hermes-Lima M (2013) Role of redox metabolism for adaptation of aquatic animals to drastic changes in oxygen availability. Comp Biochem Physiol A Mol Integr Physiol 165(4):384–404

Yadav A, Chowdhary P, Kaithwas G, Bharagava RN (2017) Toxic metals in the environment: threats on ecosystem and bioremediation approaches. In: Handbook of metal-microbe interactions and bioremediation. CRC Press, pp 128–141

Yin B, Ma H, Wang S, Chen S (2003) Electrochemical synthesis of silver nanoparticles under protection of poly (N-vinylpyrrolidone). J Phys Chem B 107(34):8898–8904

Zainith SURABHI, Sandhya S, Saxena GAURAV, Bharagava RN (2016) Microbes an ecofriendly tool for the treatment of industrial waste waters. Microbes Environ Manag 2016:75–100

Chapter 3
Emerging Water Pollutants and Their Remediation

Abstract The usage of pharmaceuticals, personal-care products, antibacterial agents, microplastics, e-waste, and nanomaterial-based products has led to a large increase in environmental pollutants. These emerging pollutants (EPs), sometimes referred to as pollutants, are causing significant concern to the general public due to substantial ecological and human well-being risks. EPs possess a wide array of ecological impacts. EPs originate from animal or human sources and have the capacity to rapidly or slowly penetrate streams via soil. As a result, drinking water sources become contaminated, water quality deteriorates, and health problems arise. Even in smaller amounts, the majority of emerging pollutants may have detrimental effects on both human health and marine ecosystems. These pollutants have no specific constraints that arise due to the capacity limitations of water treatment systems that provide drinking water and must use a variety of water sources. This degradation of aquatic habitats, specifically groundwater and the surface, is a significant issue. EP technology for treatment employs a variety of techniques, such as bioremediation, biological, contemporary oxidation processes, and physicochemical, irrespective of their advantages and constraints.

Keywords Emerging pollutants · Ecosystem · Microplastics · Nanomaterials · Groundwater · Bioremediation

3.1 Introduction

Emerging pollutants (EPs) are now a major concern for the global population due to the potential risks they pose to the environment and human health. EPs, or micropollutants, originate from a variety of sources, such as man-made and natural compounds. Food sources, surface water, municipal wastewater, groundwater, and drinking water all contain new EPs, a novel category of chemical compounds. EPs, or environmental pollutants, are chemical compounds frequently present in soil and water bodies. Only recently have we recognized them as significant water pollutants.

© The Author(s), under exclusive license to Springer Nature
Switzerland AG 2024
N. Saxena et al., *Water Pollution and Remediation*, SpringerBriefs in Water
Science and Technology, https://doi.org/10.1007/978-3-031-76301-4_3

Examples of essential pollutants commonly used and necessary in contemporary civilization include pesticides, nanomaterials, flame retardants, industrial additives, hormones, endocrine-disrupting chemicals, medications, and personal care products (Sun et al. 2024; Su et al. 2020; Birch et al. 2015; Lin et al. 2015, 2016).

According to the NORMAN network, a minimum of 700 chemicals categorised into 20 classes were detected in the aquatic environment of Europe (Peña-Guzmán et al. 2019). NORMAN network is a network of reference laboratories, research centres and related organisations for monitoring of emerging environmental substances (NORMAN n.d.). The US Geological Survey defined EP as any substance of artificial or natural origin, or any microbe not typically found in the environment, that has the potential to have adverse effects on the environment and human health. These pollutants are typically present in trace amounts ranging from parts per trillion to parts per billion (Sivaranjanee and Kumar 2021). "EPs," according to Dulio et al. (2018), are chemicals that can stay in the environment, build up in living things, and cause health problems like less fertility, abnormal growth, harm to wildlife, damage to aquatic ecosystems, neurodevelopmental delays, and possibly harm to the immune system (Rodriguez-Narvaez et al. 2017). The majority of newly emerging pollutants are not novel or have just added contaminants to the ecosystem. Newly discovered pollutants are established pollutants with recently discovered adverse effects or mechanisms of operation. Thus, the phrase "emerging" pertains to each pollutant and the resulting difficulties. We also refer to newly discovered pollutants as "contaminants" or "chemicals of emerging concern." Emerging contaminants can be categorised based on certain criteria. (i) Never consider a substance as a novel one; (ii) Identify a substance that has been in the environment for a long time but has recently gained recognition for its significance; and (iii) Identify a well-known substance whose detrimental impacts on people and the environment have only recently surfaced.

The state of the water has gotten worse because of contamination from urbanisation, agricultural operations, rapid population increases, and industrial development (Patel et al. 2019). Water quality studies generally focus on heavy metals (HMs), microbiological contaminants, prioritised contaminants, and nutrients. Recent studies (Vasilachi et al. 2021; Karpińska and Kotowska 2019) have shown the existence of organic contaminants that greatly affect water quality. When it comes to EPs, the main worry is the insufficient comprehension of their lasting impact on ecosystems, aquatic organisms, and human well-being. The identification of multiple novel substances in surface, ground, and drinking water has caused concern among people, especially in the absence of human health-based standards (Khatib et al. 2018; Baken et al. 2018). Researchers conducted several investigations to determine the levels and origins of contaminants in the bodies of water they receive (Lim et al. 2017). Different amounts and the absence of methodical observation programs limit the information available on the metabolites, transformation products, and treatment of drinking water. The majority of EPs are exempt from water and wastewater laws, resulting in a lack of knowledge about water resources.

However, governments have recently acknowledged the need to handle emerging pollutants methodically and cohesively, despite many of them still lacking regulation. The European Union (EU) has implemented intricate regulatory frameworks for EPs to regulate economic activities related to chemical contaminants in the environment. The US leads the world in both the highest limit and ongoing surveillance methods laws, thanks to its appropriate authorities (Vargas-Berrones et al. 2020).

Sutherland and Ralph's (2019) study on microalgae's potential for bioremediation of new contaminants demonstrated its capacity for concentration, filtering, removal, or biotransformation of pollutants. Gogoi et al. (2018) investigated the presence and distribution of these pollutants when treating wastewater facilities and their surrounding areas. Abaroa-Pérez et al. (2018) assessed the interaction of microplastics with other pollutants, while Khan et al. (2020) identified pharmaceuticals as a significant new contamination origin. Roy et al. (2021) focused on antibiotics as the sole new water contaminant, while Vasilachi et al. (2021) discussed surveillance, obstacles, and the adoption of effective and environmentally friendly techniques for their elimination.

This chapter provides an extensive analysis of the arrangement and origins of various emerging pollutants (EPs) derived from personal care goods, pharmaceutical companies, antibacterial agents, microplastics, electronic trash, and nanomaterials discovered worldwide. The authors also delve into various methodologies that could potentially identify EPs. The authors provide detailed discussions on various methods of remediating EPs, such as bioremediation, biological remedies, physicochemical remediation, oxidation therapies, and oxidation itself. Finally, the study examines the conclusion and future prospects.

3.2 Environmental EPs: Categorization, Sources, and Distribution

In recent times, EPs have discovered a wide range of uncontrolled contaminants in both surface and groundwater. These include antibiotic-resistant genes (ARG), antibiotic-resistant bacteria (ARB), endocrine-disrupting chemicals (EDCs), personal care products (PPCPs), pesticides, pharmaceuticals, and other chemicals and substances (Ebele et al. 2017). Many of these poisons were previously unidentified or not acknowledged until recently; however, they are now recognised as contaminants that could endanger the ecosystem. Fig. 3.1 (Mishra et al. 2023) highlights the primary sources and dispersion pathways of EPs. The potential implications for wildlife, ecosystems, animal life, and human health are receiving significant attention and require further investigation.

Bexfield et al. and Nandikes et al. have explored the long-lasting effects of particulate contaminants (EPs) on living organisms (Bexfield et al. 2019; Nandikes et al. 2022). We can classify EPs into inorganic, organic, and particulate contaminants,

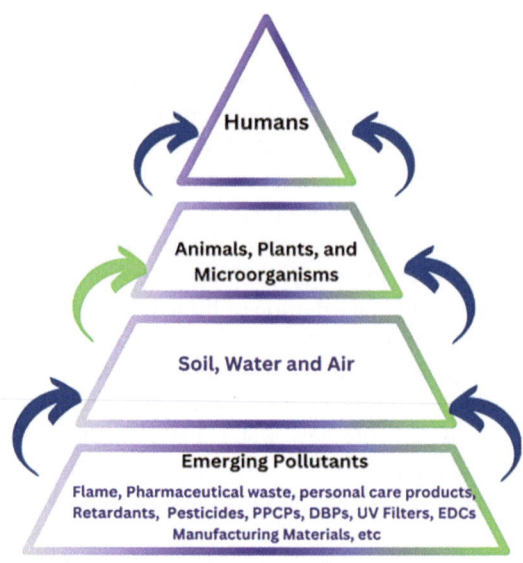

along with their respective subgroups. Table 3.1 provides a concise overview of
several groupings, subgroups, and important sources of EPs (Mishra et al. 2023).
Active pharmaceutical compounds, such as triclosan, are the primary organic molecules found in sludge, sediments, effluent, and drinking water. These compounds
originate from human and animal pharmaceuticals, containing medications, beta-
adrenergic blockers, steroids, antibiotics, and analgesics. Triclosan, an antibacterial
component present in detergents, soaps, shampoos, toothpaste, and plastic toys,
serves as the principal antiseptic (Shu et al. 2020). However, it poses a danger to
aquatic organisms and can lead to degradation products that pose significant health
hazards. Triclosan has been processed using advanced methods like oxidation,
membrane filtration, and ozonation, but their practical application is limited by
pricey processes, sludge management, and harmful byproducts (Orhon et al. 2017).
The study highlights the importance of addressing EPs in the environment to protect
living organisms and prevent their harmful effects.

EDCs, such as plasticizers, phthalates, poly-halogenates, and insecticides, can
disrupt animal growth and reproduction by mimicking hormones. Phthalates, plasticizers, poly-halogenates, and pesticides can hinder the androgen receptor's function in controlling male sexual development (Mehrpour et al. 2014). Alkyl phenols,
furans, bisphenols, heavy metals, halogenated hydrocarbons, and dioxins also hinder
estrogen's role in promoting growth in female sexuality (Kowalczyk et al. 2022).
Wastewater commonly contains normal estrogen hormones like estradiol, estriol,
and estrone, primarily from human urine (Tran et al. 2019). The body naturally
secretes estrogen consistently without the use of hormonal drugs, leading to reproductive issues in fish downstream (Marlatt et al. 2022). Pesticides, categorized as

Table 3.1 A brief summary of several types and main sources of recognised EPs

Classification of EPs	Sources	Primary pollutant	Sub-category	Refs.
Perfluorinated alkylated	Wastewater, surface water, groundwater, and sediments	Perfluorooctanoic acid and perfluorooctane sulfonat	Perfluorosulfonic and perfluorocarboxylic acids	Shahid et al. (2021)
EDCs	Surface water, drinking water, soil, sediments, and secondary sludge	Dioctyl phthalate, xenoestrogen, and bisphenol A	Phthalates, xenohormones, and bisphenol	Shahid et al. (2021)
Pharmaceutically active complexes	Wastewater from homes, livestock farms, hospitals, and health facilities, as well as from manufacturing facilities that produce medicines.	Diazepam, carbamazepine, ciprofloxacin, metoprolol, diclofenac, testosterone, and clorfibric acid	Hormones, b-blockers, lipid regulators, antibiotics, anticonvulsants, antidepressants, and nonsteroidal anti-inflammatory drugs	Shahid et al. (2021), Jacob et al. (2021)
Biocides	Aquafarming, surface water, and agricultural runoff	Epoxiconazole, butachlor, and metaldehyde	Molluscicides, fungicides, and herbicides	In Shahid et al. (2021), Rathi et al. (2021)
Regulated compounds	Surface water, soil, sewage treatment facilities, runoff from agriculture, and sediments	Phanthrene and chlorpyrifos	Poly-aromatic hydrocarbons and pesticides	Shahid et al. (2021), Jacob et al. (2020)
Surfactants	Wastewater from homes and businesses	Sodium lauryl sulfate plus tweens (Polysorbates)	Surfactants, both ionic and non-ionic	In Shahid et al. (2021), Jagini et al. (2021)
Industrial chemicals	Wastewater from both homes and businesses	Tris (1-chloro-2-propyl) phosphate with dimethyl adipate	Retardants and plasticizers	Shahid et al. (2021)
Personal care products	Surface water, landfill leachate, and WWTP effluent	Galaxolide, 4-benzophenone, Tonalide, and Diethyltoluamide	Sunblock agents/ UV filters, synthetic musk, and insect repellents	Three studies (2021): Azaroff et al. (2021), Rathi et al. (2021), and Shahid et al. (2021)

Reproduced from Ref. Mishra et al. 2023, Copyright 2023, under Creative Commons Open Access Licence from Elsevier

herbicides, insecticides, or fungicides, are significant sources of environmental pollutants in water bodies. Mojiri et al. (2020) projected the worldwide fungicide industry to reach $17.58 billion by 2022. We classify fungicides based on their chemical composition, and we classify herbicides and insecticides into different classes based on their intended use and chemical characteristics. Pesticides have a significant track record of causing harm to the ecosystem, with pesticide chemicals found in high amounts in both surface water and groundwater. Macroinvertebrates play a vital role in transporting energy to higher trophic levels in aquatic food webs, making them essential for aquatic ecosystem health (Sumudumali and Jayawardana 2021).

3.3 Kinds of EPs in the Surrounding

Arman, Shahid, and Gavrilescu have all highlighted the growing issue of contaminants in groundwater, which are not considered harmless substances (Arman et al. 2021; Shahid et al. 2021; Gavrilescu et al. 2015). Municipal trash, industrial emissions, and agricultural operations primarily cause these contaminants. Even if these unique pollutants originate from similar personal and business activities, we need specialized contamination prevention strategies (Shahid et al. 2021). People's socioeconomic and behavioural shifts may contribute to the increase in new contaminants detected, especially during the generation and transportation of drinkable water. Changes in fertilizers are important because the use of animal feces in agriculture can contaminate groundwater, cause soil erosion, and potentially cause several ailments (Arman et al. 2021). Surface water, which absorbs pollutants from sewage treatment facilities, has lower concentrations than groundwater because the wastewater gradually weakens. The concentration of contaminants in groundwater may rise if the aquifer is near affluent regions (Arman et al. 2021). The following are the primary types of EPs found in the surrounding environment.

3.3.1 EPs Produced by Personal Care Products (PCPs)

These substances are present in nutritional cosmetics, over-the-counter and prescription medications, and supplements. Wastewater treatment plants discharge these substances into the public soil and water systems. The pharmaceutical industry heavily depends on the wastewater treatment (WWT) sector (Mishra et al. 2023). Another factor contributing to PCPs in the ground is watering using contaminated river water or treated wastewater (Gallego et al. 2021). Decreased soil water affinity increases the likelihood of soil pollution by emergent contaminants. The soil is at risk of accumulating pollutants in the absence of action involving organic matter

(Beltran-Peña et al. 2020). The presence of harmful substances in areas of water where cleaned sewage is discharged is mainly due to the inadequate removal of toxins during the purification procedure, especially medicinal goods (Falahi et al. 2022).

3.3.2 EPs Produced by Pharmaceutical Industry

Despite being pollutants, researchers have found trace concentrations of pharmaceuticals in water bodies worldwide (Chinnaiyan et al. 2018). Pharmaceutical firms commonly use these chemicals in nutrition, diagnosis, treatment, and prevention, making them essential Eps (Liang et al. 2024). However, their extensive use in aquatic environments, contamination of freshwater supplies, and potential damage to biodiversity and humans make them essential EPs (Sharma et al. 2019). The introduction of pharmaceuticals into water bodies can mitigate environmental risks by diluting sewage (Patel et al. 2019). Farm animals, cattle, and fisheries are among the many uses for medications, with around 3000 chemicals in pharmaceutical goods. Researchers have investigated the potential harm of ng L^{-1} doses in the air to humans and other forms of biodiversity (Mishra et al. 2023). The majority of medicines released into the atmosphere come from human urine, the disposal of medication waste, and agricultural uses. Pharmaceuticals can have a knock-on effect on the environment through organic fertilizers used in intensive animal husbandry, such as dung and purines. They may enter the food chain and infect all forms of life. Tahira et al. (2022) discovered that wastewater and drinking water commonly contain medications such as antibiotics, antidepressants, ciprofloxacin, antacids, clofibric acid, beta blockers, fluoxetine, nitroglycerin, tranquilizers, stimulants, anti-inflammatory pharmaceuticals, propranolol, antipyretics, cholesterol-lowering drugs, and analgesics.

3.3.3 EPs Produced by Antibacterial Agents

One common antiseptic is triclosan, which is present in many home items, such as clothing, soap, and toothpaste. In tap water, triclosan combines with chlorine to produce chloroform as a secondary substance. Triclosan, or methyl triclosan, breaks down into various dioxins in the environment (Gwenzi et al. 2018). Triclosan is incredibly toxic to certain species of earthworms and algae. Earthworms undergo oxidative stress, leading to changes in their metabolism and significant DNA damage (Gillis et al. 2017). We haven't determined the water security criteria for triclosan. Fluoroquinolone is another antibacterial compound in use. Both humans and animals have historically used fluoroquinolone antibiotics for treatment. There are around 60 fluoroquinolones, with approximately 20 of them having authorized arrays. Fluoroquinolones have the potential to harm a range of microorganisms and cause harm to the environment. Fluoroquinolones can have adverse effects on soil and groundwater, even in very small amounts (Chen et al. 2015). In recent years, there has been a significant increase in studies focused on removing antimicrobials and novel contaminants. The number of papers addressing this subject has notably risen from 2010 to 2021 (Palacio et al. 2022).

3.3.4   EPs Produced by Microplastics

The annual production of plastic has grown significantly since the 1950s, resulting in 359 million metric tonnes of new plastic (Geyer 2020). This growth has led to the introduction of microplastics, small plastic fragments less than 5 mm in length, into the environment. These fragments pose a significant environmental problem, as they can be hazardous and transport other toxins (Zhao et al. 2024; Stapleton and Hai 2023). Microplastics negatively impact marine organisms' ingestion rates and growth, and they can also impede plant roots and leaves (Cole et al. 2015; Stapleton and Hai 2023). Contaminated microplastics can transfer the pollutant to host organisms. Despite the uncertainty surrounding the scope of microplastic problems, recent research has confirmed their classification as a substantial and worrisome pollutant (Stapleton and Hai 2023).

Environmental experts categorize microplastic pollution into primary and secondary sources (Duis and Coors 2016). Primary microplastics are intentionally produced microplastics of a size of 5 mm or less, while secondary sources include disintegration due to weathering or abrasive forces and tyre production on roadways (Faure et al. 2015). The identification of microplastics' sources varies depending on the specific literature being studied. For example, stormwater runoff and sewage outflow may be the primary causes of microplastic pollution in oceans or rivers, but the exact mechanism by which microplastics enter these systems remains incompletely understood. Identifying the primary sources of microplastic pollution has become a significant area of study in recent publications, providing a clear description of well-known and newly identified sources of microplastic pollution. A simplified elucidation of the well-established and newly identified origins of microplastic pollution, as illustrated in Fig. 3.2 (Stapleton and Hai 2023).

Microplastic contamination has been recorded in land, water, and air settings (Strady et al. 2021; Zhang et al. 2021). Microplastics have inevitably been exposed in the human body due to the trophic transmission of microplastics in the aquatic food chain (Tang et al. 2021). The dissemination of microplastics in the environment is substantial, and the impact of microplastic pollution on humans, plants, and animals is a specific area of interest in current research. The primary focus of concern with microplastic contamination includes (Stapleton and Hai 2023): (i) The impact of microplastics on humans, (ii) The impact of microplastics on wildlife, and (iii) The impact of microplastics on plant life. For additional information, please see reference (Stapleton and Hai 2023).

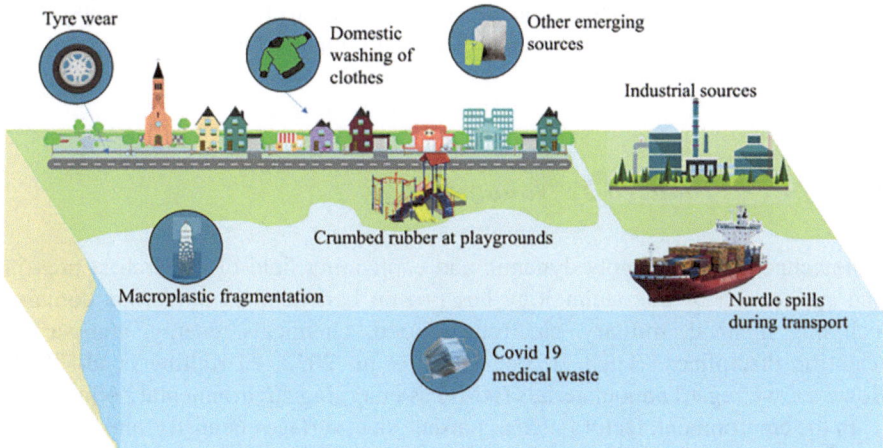

Fig. 3.2 Existing and potential future causes of microplastic contamination. (Reproduced from Ref. Stapleton and Hai 2023, Copyright 2023, under Creative Commons Open Access Licence from Taylor & Francis)

3.3.5 EPs Produced by E-Wastes

Electronic waste, particularly mobile phones, is a significant global issue due to the electronics industry's rapid growth. Arya and Kumar (2020) highlight the global demand and consumption rates for these devices, which contribute significantly to the global economy. However, only 5 billion people have the means to meet their fundamental needs, and over 6 billion people are smartphone users, which has become an essential requirement (Krishnamoorthy et al. 2018; Jain et al. 2023). This waste poses a threat to human health and the environment. The increasing demand for electronic waste, such as laptops, computers, solar panels, electric cars, and mobile phones, has led to significant growth in the IT sector, resulting in a strong economic performance for countries. Domestic e-waste is generated by home appliances, small and large enterprises, PC manufacturers, establishments, and other sectors. However, many countries prohibit the disposal of such devices, leading to complications in the management of e-waste. Factors such as reduced processing costs, decreased personnel expenses, and inadequate enforcement of environmental regulations are responsible for the increase in illegal e-waste imports. The lack of reliable data on e-waste production, imports, and exports makes it impossible to accurately determine its quantity.

Both people and the environment experience a range of detrimental effects caused by electrical waste. As a result of inadequate management and technological improvements, this scenario has evolved into a significant threat to human life. Electronic trash has a detrimental impact on both the environment and human health. The primary detrimental impacts of e-waste (Jain et al. 2023) include: (i) adverse effects on human health; (ii) detrimental effects on the environment;

(iii) harmful consequences for air quality; (iv) detrimental effects on soil composition; and (v) adverse impacts on water quality. For additional information, please refer to reference (Jain et al. 2023).

3.3.6 EPs Produced by Nanomaterials

Nanotechnology is a highly dynamic and captivating field of research at present. The application of this technology has proven beneficial in numerous domains, including medical, military, electronics, food, chemicals, energy, and various scientific disciplines (Saleh 2020a; Souza et al. 2022; El-Kalliny et al. 2023). However, we regard nanomaterials (NMs) as emerging environmental pollutants.

In the environment, factors such as particle size, surface chemistry, and both non-living and living processes influence the dispersion and movement of nanomaterials (NMs), including nanoparticles (NPs). As a result, NPs can either exist as individual particles in suspension or combine to form larger NMs. It has the potential to exist in a dissolved state or undergo a reaction with natural elements (Saleh 2020b). The minuscule dimensions of NPs result in a sluggish rate of sedimentation, leading to prolonged suspension in both water and air. Thus, smaller NPs exhibit greater ease and distance of transfer compared to larger ones made of the same substance (El-Kalliny et al. 2023). Rainwater runoff or wind, either intentionally or unintentionally, releases NMs from solid waste and wastewater effluents into the aquatic environment (Klaine et al. 2008; Lowry et al. 2010). The smaller dimensions of NMs have a greater impact on their ability to move across porous environments, facilitating strong adhesion and agglomeration on mineral surfaces. For example, the binding of nanoparticles to mineral surfaces hinders their movement in groundwater aquifers (Wiesner et al. 2006). Consequently, the soil traps nanoparticles (NMs) after they travel longer distances. Soils that contain a lot of clay have the ability to stabilize NMs, which allows for greater dissemination (El-Kalliny et al. 2023). To obtain additional information, please consult El-Kalliny et al. (2023).

3.4   Techniques for Identifying EPs

Developing reliable and validated quantitative methods and tools to detect novel contaminants for global laboratory use is a key challenge for the chemical technology field. It is critical to analyse the identification and chromatographic separation of new organic pollutants in a variety of sturdy environmental materials. Prior to the examination, several other crucial treatment steps are frequently necessary, such as filtering, sample concentration, framework segregation, and pH correction (Rathi et al. 2021). To analyse EPs, sample preparation and instrumentation evaluation are required.

3.4.1   Preparation of the Sample

Solid-phase extraction, ultrasound-assisted extraction, membrane technologies, and derivatization techniques are some of the analytical methods that Rathi, Jothirani, and Mishra have looked into the preparation (Jothirani et al. 2016). The widespread use of solid-phase extraction stems from its simplicity, reliability, and accuracy. Molecularly imprinted polymers, digital solid phase extraction, automatic solid phase extraction, stirring sorptive bar extraction, solid-state microextraction, and magnetic solid phase extraction are some other ways to do it. Membrane technologies like reverse osmosis and ultrafiltration separate larger organic compounds from smaller water molecules.

3.4.2   Instrumental Assessment

Chromatography is a crucial tool for identifying emerging pollutants by analyzing their polarisation, volatility, and thermal characteristics in both liquid and gas forms (Hernández et al. 2015). With more than a thousand novel pollutants currently identified, it can be difficult to use conventional gas chromatography-mass spectrometry (GC-MS) to trace them (Lebedev et al. 2020). Liquid chromatography-mass spectrometry has significantly advanced, making it a crucial technology for evaluating emerging contaminants. Researchers have widely used high-tech instruments like liquid chromatography-mass spectrophotometers, liquid chromatography-tandem mass spectroscopy, gas chromatography-mass spectrometry, and gas chromatography tandem mass spectrometry to research EPs in aqueous ecosystems (Wu et al. 2010). The most widely utilized mobile phase in supercritical fluid chromatography is carbon dioxide due to its safety, non-explosive nature, and simple laboratory conditions. Supercritical-fluid chromatography identified seven emerging pollutants, including pharmaceuticals, endocrine disruptors, bactericides, and pesticides.

   Snow et al. (2015) developed a new analytical technique that simultaneously identified ten different steroid hormones in agricultural, animal, or dirt dung using tandem mass spectrometry, gas chromatography, and its reversed and combined high-pressure fluid removal counterparts. This method uses liquid chromatography, tandem mass spectrometry, and electrospray to identify 26 animal-healthy penicillins in pig effluent. Zonja et al. (2014) used quadrupole linear ion trap mass spectrometry and high-performance liquid chromatography to find ten different diabetes medicines in wastewater. Improved quadruple time-of-flight and orbitrap mass spectroscopy are two types of high-resolution mass spectrometry instruments that have made significant progress in the lab-scale analysis of organic chemicals that are becoming pollutants. Richardson & Kimura (2019) suggest that employing immuno-analytical techniques may prove beneficial in the trace management of organic environmental pollutants, as they are less expensive, require little sample preparation, and demonstrate exceptional precision. However, antibiotic selectivity prevents their application in immunoassays that aim to test compounds from multiple classes at once.

3.5 Approaches to Remediating EPs

Mohapatra & Kirpalani, Rosal, and other researchers have highlighted the importance of addressing emerging pollutants in wastewater treatment plants (Mohapatra and Kirpalani 2019; Rosal et al. 2010). These pollutants can have harmful effects on humans, such as bacterial resistance, feminization of aqueous organisms, neurotoxic effects, disruption of the endocrine system, and tumor formation (Verma et al. 2024). Conventional wastewater treatment technologies employ procedures using physical, chemical, and biological methods to eliminate both soluble and insoluble contaminants. Researchers have developed various techniques to eliminate EPs, such as adsorption, membrane technology, accelerated oxidation processes, biological techniques, and wetlands. The octanol-water distribution coefficient (D_{ow}) and the octanol-water partition coefficient (Kow) largely determine the hydrophobic properties of the pollutants. Positive Log (K_{ow}) values indicate hydrophobicity, while lower values indicate more wettability. Rosal et al. (2010) highlighted the slow dissociation of basic or acidic components in polar compounds. Figure 3.3 shows different treatment methods for removing EPs and their consumers.

3.5.1 Bioremediation

Bioremediation is the controlled biological breakdown of organic wastes (Sun et al. 2024; Malik et al. 2022; Alvarez et al. 2017; Priyadarshanee and Das 2021). Malik, Alvarez, Priyadarshanee, and Das have all made progress in this field. Enzyme activity is crucial for each phase of the metabolic cycle, and living things

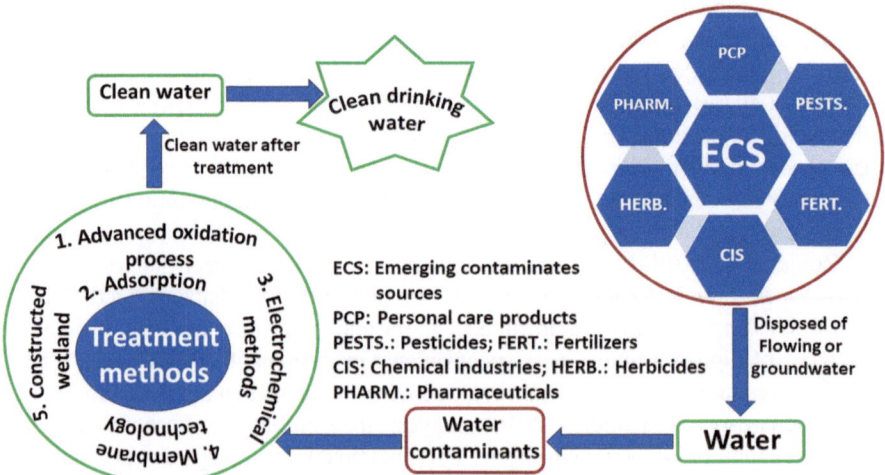

Fig. 3.3 Various treatment methods for the removal of EPc and their applications. (Reproduced from Ref. Mishra et al. 2023, Copyright 2023, under Creative Commons Open Access Licence from Elsevier)

and fertilizers can assist in the natural bioremediation process. Biodegradation, a crucial component of bioremediation technology, transforms harmful organic pollutants into safe inorganic compounds that aquatic life, plants, animals, and people can utilize. Bioremediation employs microorganisms, such as bacteria, algae, fungi, and plants, to break down, transform, eliminate, freeze, or sanitize environmental toxins. Environmental, chemical, biological, soil type, carbon and nitrogen sources, and microbe type all have an impact on the bioremediation process. Carbon is an important part of in situ bioremediation because it speeds up the breakdown of pollutants and boosts the metabolism of native bacterial species. Bioremediation employs bacteria, microorganisms, and fungi to eliminate environmental pollutants from soil and groundwater. Figure 3.4 shows the bioremediation of emerging contaminants.

3.5.2   Biological Remedies

Since the twentieth century, wastewater has used conventional activated sludge (CAS), a conventional biological treatment method, to decompose organic matter, contaminants, and nutrients. However, its use in drinking water is less frequent.

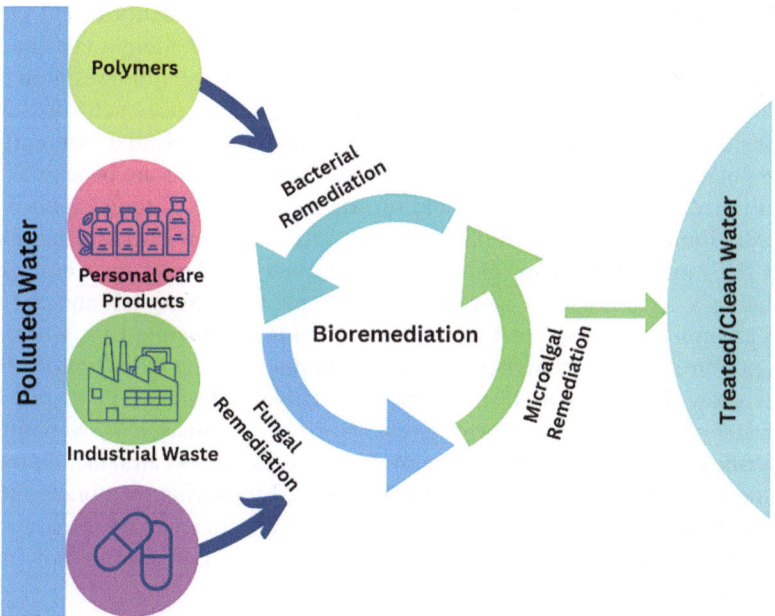

Fig. 3.4 Schematic diagram illustrating the process of bioremediation for emerging contaminants. (Reproduced from Ref. Mishra et al. 2023, Copyright 2023, under Creative Commons Open Access Licence from Elsevier)

Advancements in CAS are enhancing its acceptance, feasibility, and effectiveness. Hussain and Dubey (2014) have highlighted its limitations, such as lengthy hydraulic retention times and higher operating requirements. Despite its environmental benefits, CAS has drawbacks such as increased sludge production, high energy consumption, and potential accumulation in secondary clarifiers. Despite these challenges, CAS remains an economical and environmentally friendly substitute for conventional sewage treatment.

Membrane bioreactors, or MBRs, are a biological treatment that might work well for getting rid of organic matter in wastewater. They work very well in facilities that make drugs, chemical oxygen demand (COD), and biological oxygen demand (BOD) (Mishra et al. 2023). According to Sathishkumar et al. (2014), treating laccase with syringaldehyde effectively reduced the harmful effects of diclofenac on cells. Syringaldehyde and laccase convert triclosan's dichlorination product into 2-phenoxyphenol, a harmless polymer. When combined with laccase-poly nanofibers, syringaldehyde can effectively remove diclofenac from water sources. Due to their environmental friendliness, enzymes have the potential to treat environmental pollutants.

3.5.3 Physico-Chemical Remediation

Conventional water treatment methods frequently involve chemical and physical methods to remove infectious agents, address turbidity, and address taste and odor issues. However, these methods are frequently ineffective and often result in micropollutants in drinking water. The adsorption process now uses activated carbon, such as powdered activated carbon (PAC) and granulated activated carbon (GAC), to treat potable water and industrial effluent. Activated carbons are beneficial due to their advanced chemistry, porosity, and surface area. However, their expensive cost frequently limits their use (Nazifa et al. 2018). Reverse osmosis (RO) and nanofiltration (NF) membranes have gained significant interest due to their efficiency in eliminating a wide range of inorganic and organic pollutants. NF and RO membranes are more effective than nanofiltration (NF) in eliminating EPs, whereas RO is less preferred due to its higher energy consumption during the filtration process. Studies show that RO and NF membranes effectively eliminate PCPs and EDCs from water plant treatment, achieving removal rates as high as 95%. Nanofiltration in water-based systems effectively removes pesticides such as atrazine, diuron, and simazine, along with the atrazine metabolite DEA. Employing membrane filtration as the final stage can enhance nanoparticle removal efficiency. However, fouling hinders the membrane filtering approach, leading to a decrease in flow and an increase in operational costs (Liu et al. 2014).

3.5.4   Oxidation Treatment

The process of oxidation is a crucial technique for addressing newly identified contaminants by utilising chemical oxidants like ozone or chlorine. Water reactions can be quite reactive, resulting in secondary products. Therefore, it is important to carefully choose chemical oxidants before opting for this treatment. Ozone (O_3) interacts with organic contaminants to generate molecular O_3 and radicals in nature through breakdown (Saravanan et al. 2021). More hydroxyl radicals are made and carbon-halogen bonds are broken with photons in advanced oxidation processes (AOPs) based on O_3. This makes them more effective than ozonation alone. Examples of these AOPs are O_3/UV, photochemical, Fenton-type techniques, and O_3/H_2O_2. AOPs are the preferred method for treating stubborn chemicals (Wang and Zhuan 2020). UV photolysis has effectively destroyed organophosphorus pesticides utilising the Lamps for Xe, Hg, and Sunset devices, resulting in different mechanisms, kinetics, and by-product formation. In some cases, UV photolysis under low pressure was able to break down pesticides like pentachlorophenol, diuron, atrazine, and alachlorp, even when very toxic oxons were present (Li et al. 2019). A direct photolysis method was only able to partially degrade pesticides in a water-based solution. When combined with H_2O_2 or Fe (III), these light sources demonstrated enhanced performance compared to photolysis alone (Shawaqfeh and Al Momani 2010). Researchers commonly study TiO_2, a semiconductor catalyst, using a variety of photocatalytic systems for increased oxidation treatments. Titanium dioxide (TiO_2) possesses the benefits of non-toxicity, chemical stability, and cost-effectiveness. TiO_2 can serve as a photocatalyst for purifying effluent by photocatalytic treatment. Nevertheless, the problem is that it is a powder, necessitating a segregation process after purification. TiO_2 has outstanding photocatalytic efficiency in removing paracetamol using wastewater to extract it, achieving a 99–100% removal rate within 4 h of exposure to light (Borges et al. 2015).

3.5.5   Oxidation

Adsorption is a crucial technique for separating low-concentration contaminants, as demonstrated by several works (Mishra et al. 2023; Rathi et al. 2021). Adsorption is effective and reliable in treating wastewater due to its simplicity, versatility, and lack of reliance on harmful chemicals (Rathi et al. 2021). Scientists have conducted research to address emerging contaminants found in sludge from water and wastewater systems, wastewater from the paper industry, silt, soil, fertilizers, and medications in the environment. Components used in sorbents include nanocomposites, mesoporous clay nanocomposites, carbon nanotubes, graphene, activated carbon, hydrochar, biochar, and carbon composite materials. Researchers have used nanocomposites to eliminate emerging pollutants from wastewater releases (Lima et al. 2019). Antibiotics and other organic substances are removed by activated carbon through non-specific scattering interactions. Ionic or polar antibiotics are removed by electrostatic contact with the activated carbon's surface charge group. Researchers have used new pollutants like bisphenol-A and activated carbon powder to eliminate luoroquinolonic acids, caffeine, clofibric acid, diclofenac, gallic acid, ibuprofen, salicylic acid, and ofloxacin (Lofrano 2012). Because it has more surface area and smaller pores, biochar was better at getting rid of benzophenol, bisphenol, benzotriazole, and 17-estradiol by 5–30% compared to powdered activated carbon. SSA, functional groups, pore volume, dispersion, and surface charge are some of the surface properties of changed biochar that are important for EP adsorption. Changing the chemical structure of biochar can improve its ability to absorb things by creating more adsorption sites and making it more open to electrostatic attraction, surface precipitation, and surface complexation (Rajapaksha et al. 2016). Researchers identify activated carbon's biochar as a promising material for the adsorption of EPs like heavy metals and fertilizers. However, elements such as surface charge, pore volume, and surface area limit its widespread use.

3.6   Conclusion and Future Perspectives

EPs are abundant and pose a significant risk to human health and the ecosystem, even in small quantities. Anthropogenic and naturally occurring substances, which are not commonly regarded as pollutants, along with novel materials generated and utilised by humanity, have been significantly modifying the marine ecology. Further investigation is necessary should choose appropriate methods for investigation, or to remove pollutants at the very least using established treatment methods. Environmental studies now encompass a wider range of substances, such as pharmaceuticals, microplastics, e-waste, nanomaterials, and items for personal hygiene used at home. Researchers are investigating these substances apart from the conventional hazardous substances that have been the focus of long-standing research. EPs place a high priority on pharmaceutical chemicals because of their significant impact on microbial systems, human well-being, aquatic and land environments, and the overall environment. Improving energy efficiency is essential for reducing energy use and minimising environmental effects. Future research might prioritise the evolution of novel construction materials, energy management technologies, cutting-edge transit systems, and energy-saving appliances to enhance energy efficiency. Further investigation could centre regarding methods for managing demand, including as time-of-use pricing, behavioural interventions, and automated demand response.

References

Abaroa-Pérez B, Sánchez-Almeida G, Hernández-Brito JJ, Vega-Moreno D (2018) In situ miniaturised solid phase extraction (m-SPE) for organic pollutants in seawater samples. J Anal Methods Chem 2018. https://doi.org/10.1155/2018/7437031

Alvarez A, Saez JM, Costa JSD, Colin VL, Fuentes MS, Cuozzo SA et al (2017) Actinobacteria: current research and perspectives for bioremediation of pesticides and heavy metals. Chemosphere 166:41–62. https://doi.org/10.1016/j.chemosphere.2016.09.070

Arman NZ, Salmiati S, Aris A, Salim MR, Nazifa TH, Muhamad MS, Marpongahtun M (2021) A review on emerging pollutants in the water environment: existences, health effects and treatment processes. Water 13(22):3258. https://doi.org/10.3390/w13223258

Arya S, Kumar S (2020) E-waste in India at a glance: current trends, regulations, challenges and management strategies. J Clean Prod 271:122707. https://doi.org/10.1016/j.jclepro.2020.122707

Azaroff A, Monperrus M, Miossec C, Gassie C, Guyoneaud R (2021) Microbial degradation of hydrophobic emerging contaminants from marine sediment slurries (Capbreton canyon) to pure bacterial strain. J Hazard Mater 402:123477. https://doi.org/10.1016/j.jhazmat.2020.123477

Baken KA, Sjerps RM, Schriks M, van Wezel AP (2018) Toxicological risk assessment and prioritization of drinking water relevant contaminants of emerging concern. Environ Int 118:293–303. https://doi.org/10.1016/j.envint.2018.05.006

Beltran-Peña A, Rosa L, D'Odorico P (2020) Global food self-sufficiency in the 21st century under sustainable intensification of agriculture. Environ Res Lett 15(9):095004

Bexfield LM, Toccalino PL, Belitz K, Foreman WT, Furlong ET (2019) Hormones and pharmaceuticals in groundwater used as a source of drinking water across the United States. Environ Sci Technol 53(6):2950–2960. https://doi.org/10.1021/acs.est.8b05592

Birch GF, Drage DS, Thompson K, Eaglesham G, Mueller JF (2015) Emerging contaminants (pharmaceuticals, personal care products, a food additive and pesticides) in waters of Sydney estuary, Australia. Mar Pollut Bull 97(1–2):56–66. https://doi.org/10.1016/j.marpolbul.2015.06.038

Borges ME, García DM, Hernández T, Ruiz-Morales JC, Esparza P (2015) Supported photocatalyst for removal of emerging contaminants from wastewater in a continuous packed-bed photoreactor configuration. Catalysts 5(1):77–87. https://doi.org/10.3390/catal5010077

Chen G, Li M, Liu X (2015) Fluoroquinolone antibacterial agent contaminants in soil/groundwater: a literature review of sources, fate, and occurrence. Water Air Soil Pollut 226:1–11. https://doi.org/10.1007/s11270-015-2438-y

Chinnaiyan P, Thampi SG, Kumar M, Mini K (2018) Pharmaceutical products as emerging contaminant in water: relevance for developing nations and identification of critical compounds for Indian environment. Environ Monit Assess 190:1–13. https://doi.org/10.1007/s10661-018-6672-9

Cole M, Lindeque P, Fileman E, Halsband C, Galloway TS (2015) The impact of polystyrene microplastics on feeding, function and fecundity in the marine copepod Calanus helgolandicus. Environ Sci Technol 49(2):1130–1137. https://doi.org/10.1021/es504525u

Duis K, Coors A (2016) Microplastics in the aquatic and terrestrial environment: sources (with a specific focus on personal care products), fate and effects. Environ Sci Eur 28(1):2. https://doi.org/10.1186/s12302-015-0069-y

Dulio V, van Bavel B, Brorström-Lundén E, Harmsen J, Hollender J, Schlabach M et al (2018) Emerging pollutants in the EU: 10 years of NORMAN in support of environmental policies and regulations. Environ Sci Eur 30(1):1–13. https://doi.org/10.1186/s12302-018-0135-3

Ebele AJ, Abdallah MAE, Harrad S (2017) Pharmaceuticals and personal care products (PPCPs) in the freshwater aquatic environment. Emerging Contaminants 3(1):1–16. https://doi.org/10.1016/j.emcon.2016.12.004

El-Kalliny AS, Abdel-Wahed MS, El-Zahhar AA, Hamza IA, Gad-Allah TA (2023) Nanomaterials: a review of emerging contaminants with potential health or environmental impact. Discover Nano 18(1):68. https://doi.org/10.1186/s11671-023-03787-8

Falahi O A A, Abdullah S R S, Hasan H A, Othman A R, Ewadh H M, Kurniawan S B, Imron M F (2022) Occurrence of pharmaceuticals and personal care products in domestic wastewater, available treatment technologies, and potential treatment using constructed wetland: a review. Process Saf Environ Prot 168:1067–1088

Faure F, Demars C, Wieser O, Kunz M, De Alencastro LF (2015) Plastic pollution in Swiss surface waters: nature and concentrations, interaction with pollutants. Environ Chem 12(5):582–591. https://doi.org/10.1071/EN14218

Gallego S, Montemurro N, Béguet J, Rouard N, Philippot L, Pérez S, Martin-Laurent F (2021) Ecotoxicological risk assessment of wastewater irrigation on soil microorganisms: fate and impact of wastewaterborne micropollutants in lettuce-soil system. Ecotoxicol Environ Saf 223:112595

Gavrilescu M, Demnerová K, Aamand J, Agathos S, Fava F (2015) Emerging pollutants in the environment: present and future challenges in biomonitoring, ecological risks and bioremediation. New Biotechnol 32(1):147–156. https://doi.org/10.1016/j.nbt.2014.01.001

Geyer R (2020) Chapter 2-production, use, and fate of synthetic polymers. In: Letcher TM (ed) Plastic waste and recycling, pp 13–32. https://doi.org/10.1016/B978-0-12-817880-5.00002-5

Gillis JD, Price GW, Prasher S (2017) Lethal and sub-lethal effects of triclosan toxicity to the earthworm Eisenia fetida assessed through GC–MS metabolomics. J Hazard Mater 323:203–211. https://doi.org/10.1016/j.jhazmat.2016.07.022

Gogoi A, Mazumder P, Tyagi VK, Chaminda GT, An AK, Kumar M (2018) Occurrence and fate of emerging contaminants in water environment: a review. Groundw Sustain Dev 6:169–180. https://doi.org/10.1016/j.gsd.2017.12.009

Gwenzi W, Mangori L, Danha C, Chaukura N, Dunjana N, Sanganyado E (2018) Sources, behaviour, and environmental and human health risks of high-technology rare earth elements as emerging contaminants. Sci Total Environ 636:299–313. https://doi.org/10.1016/j.scitotenv.2018.04.235

Hernández F, Ibáñez M, Portolés T, Cervera MI, Sancho JV, López FJ (2015) Advancing towards universal screening for organic pollutants in waters. J Hazard Mater 282:86–95. https://doi.org/10.1016/j.jhazmat.2014.08.006

Hussain A, Dubey SK (2014) Specific methanogenic activity test for anaerobic treatment of phenolic wastewater. Desalin Water Treat 52(37–39):7015–7025. https://doi.org/10.1080/19443994.2013.823116

Jacob MM, Ponnuchamy M, Kapoor A, Sivaraman P (2020) Bagasse based biochar for the adsorptive removal of chlorpyrifos from contaminated water. J Environ Chem Eng 8(4):103904. https://doi.org/10.1016/j.jece.2020.103904

Jacob RS, Araújo CV, de Souza Santos LV, Moreira VR, Lebron YAR, Lange LC (2021) The environmental risks of pharmaceuticals beyond traditional toxic effects: chemical differences that can repel or entrap aquatic organisms. Environ Pollut 268:115902. https://doi.org/10.1016/j.envpol.2020.115902

Jagini S, Thaduri S, Konda S, Saranga VK, Dheeravath B, Vurimindi H (2021) Emerging contaminant (Triclosan) removal by adsorption and oxidation process: comparative study. Model Earth Syst Environ 7:2431–2438. https://doi.org/10.1007/s40808-020-01020-4

Jain M, Kumar D, Chaudhary J, Kumar S, Sharma S, Verma AS (2023) Review on E-waste management and its impact on the environment and society. Waste Manag Bull. https://doi.org/10.1016/j.wmb.2023.06.004

Jothirani R, Kumar PS, Saravanan A, Narayan AS, Dutta A (2016) Ultrasonic modified corn pith for the sequestration of dye from aqueous solution. J Ind Eng Chem 39:162–175. https://doi.org/10.1016/j.jiec.2016.05.024

Karpińska J, Kotowska U (2019) Removal of organic pollution in the water environment. Water 11(10):2017. https://doi.org/10.3390/w11102017

Khan NA, Khan SU, Ahmed S, Farooqi IH, Yousefi M, Mohammadi AA, Changani F (2020) Recent trends in disposal and treatment technologies of emerging-pollutants-a critical review. TrAC Trends Anal Chem 122:115744

Khatib JM, Baydoun S, ElKordi AA (2018) Water pollution and urbanisation trends in Lebanon: Litani River basin case study. Urban Pollut Sci Manag:397–415. https://doi.org/10.1002/9781119260493.ch30

Klaine SJ, Alvarez PJ, Batley GE, Fernandes TF, Handy RD, Lyon DY et al (2008) Nanomaterials in the environment: behavior, fate, bioavailability, and effects. Environ Toxicol Chem Int J 27(9):1825–1851. https://doi.org/10.1897/08-090.1

Kowalczyk A, Wrzecińska M, Czerniawska-Piątkowska E, Araújo JP, Cwynar P (2022) Molecular consequences of the exposure to toxic substances for the endocrine system of females. Biomed Pharmacother 155:113730. https://doi.org/10.1016/j.biopha.2022.113730

Krishnamoorthy Y, Sakthivel M, Sarveswaran G (2018) Emerging public health threat of e-waste management: global and Indian perspective. Rev Environ Health 33(4):321–329. https://doi.org/10.1515/reveh-2018-0021

Lebedev AT, Mazur DM, Artaev VB, Tikhonov GY (2020) Better screening of non-target pollutants in complex samples using advanced chromatographic and mass spectrometric techniques. Environ Chem Lett 18:1753–1760. https://doi.org/10.1007/s10311-020-01037-2

Li W, Zhao Y, Yan X, Duan J, Saint CP, Beecham S (2019) Transformation pathway and toxicity assessment of malathion in aqueous solution during UV photolysis and photocatalysis. Chemosphere 234:204–214. https://doi.org/10.1016/j.chemosphere.2019.06.058

Liang S, Wang X, Li C, Liu L (2024) Biological activity of lactic acid bacteria exopolysaccharides and their applications in the food and pharmaceutical industries. Food Secur 13(11):1621. https://doi.org/10.3390/foods13111621

Lim FY, Ong SL, Hu J (2017) Recent advances in the use of chemical markers for tracing wastewater contamination in aquatic environment: a review. Water 9(2):143. https://doi.org/10.3390/w9020143

Lima DR, Gomes AA, Lima EC, Umpierres CS, Thue PS, Panzenhagen JC et al (2019) Evaluation of efficiency and selectivity in the sorption process assisted by chemometric approaches: removal of emerging contaminants from water. Spectrochim Acta A Mol Biomol Spectrosc 218:366–373. https://doi.org/10.1016/j.saa.2019.04.018

Lin YC, Lai WWP, Tung HH, Lin AYC (2015) Occurrence of pharmaceuticals, hormones, and perfluorinated compounds in groundwater in Taiwan. Environ Monit Assess 187:256. https://doi.org/10.1007/s10661-015-4497-3

Lin T, Yu S, Chen W (2016) Occurrence, removal and risk assessment of pharmaceutical and personal care products (PPCPs) in an advanced drinking water treatment plant (ADWTP) around Taihu Lake in China. Chemosphere 152:1–9. https://doi.org/10.1016/j.chemosphere.2016.02.109

Liu P, Zhang H, Feng Y, Yang F, Zhang J (2014) Removal of trace antibiotics from wastewater: a systematic study of nanofiltration combined with ozone-based advanced oxidation processes. Chem Eng J 240:211–220. https://doi.org/10.1016/j.cej.2013.11.057

Lofrano G (ed) (2012) Emerging compounds removal from wastewater: natural and solar based treatments. Springer Science & Business Media

Lowry GV, Hotze EM, Bernhardt ES, Dionysiou DD, Pedersen JA, Wiesner MR, Xing B (2010) Environmental occurrences, behavior, fate, and ecological effects of nanomaterials: an introduction to the special series. J Environ Qual 39(6):1867–1874. https://doi.org/10.2134/jeq2010.0297

Malik JA, Goyal MR, Wani KA (2022) Bioremediation and phytoremediation technologies in sustainable soil management: volume 4: degradation of pesticides and polychlorinated biphenyls. CRC Press

Marlatt VL, Bayen S, Castaneda-Cortès D, Delbès G, Grigorova P, Langlois VS et al (2022) Impacts of endocrine disrupting chemicals on reproduction in wildlife and humans. Environ Res 208:112584. https://doi.org/10.1016/j.envres.2021.112584

Mehrpour O, Karrari P, Zamani N, Tsatsakis AM, Abdollahi M (2014) Occupational exposure to pesticides and consequences on male semen and fertility: a review. Toxicol Lett 230(2):146–156. https://doi.org/10.1016/j.toxlet.2014.01.029

Mishra RK, Mentha SS, Misra Y, Dwivedi N (2023) Emerging pollutants of severe environmental concern in water and wastewater: a comprehensive review on current developments and future research. Water-Energy Nexus 6:74–95. https://doi.org/10.1016/j.wen.2023.08.002

Mohapatra DP, Kirpalani DM (2019) Advancement in treatment of wastewater: fate of emerging contaminants. Can J Chem Eng 97(10):2621–2631. https://doi.org/10.1002/cjce.23533

Mojiri A, Zhou JL, Robinson B, Ohashi A, Ozaki N, Kindaichi T et al (2020) Pesticides in aquatic environments and their removal by adsorption methods. Chemosphere 253:126646. https://doi.org/10.1016/j.chemosphere.2020.126646

Nandikes G, Pathak P, Razak AS, Narayanamurthy V, Singh L (2022) Occurrence, environmental risks and biological remediation mechanisms of Triclosan in wastewaters: challenges and perspectives. J Water Process Eng 49:103078. https://doi.org/10.1016/j.jwpe.2022.103078

Nazifa TH, Habba N, Salmiati A, Hadibarata T (2018) Adsorption of Procion red MX-5B and crystal violet dyes from aqueous solution onto corncob activated carbon. J Chin Chem Soc 65(2):259–270. https://doi.org/10.1002/jccs.201700242

NORMAN (n.d.). https://www.norman-network.net/

Orhon KB, Orhon AK, Dilek FB, Yetis U (2017) Triclosan removal from surface water by ozonation-kinetics and by-products formation. J Environ Manag 204:327–336. https://doi.org/10.1016/j.jenvman.2017.09.025

Palacio DA, Aranda FL, Rivas BL (2022) Removal of antibiotic emerging pollutants: an overview. J Chil Chem Soc 67(3):5547–5561. https://doi.org/10.4067/S0717-97072022000305547

Patel M, Kumar R, Kishor K, Mlsna T, Pittman CU Jr, Mohan D (2019) Pharmaceuticals of emerging concern in aquatic systems: chemistry, occurrence, effects, and removal methods. Chem Rev 119(6):3510–3673. https://doi.org/10.1021/acs.chemrev.8b00299

Peña-Guzmán C, Ulloa-Sánchez S, Mora K, Helena-Bustos R, Lopez-Barrera E, Alvarez J, Rodriguez-Pinzón M (2019) Emerging pollutants in the urban water cycle in Latin America: a review of the current literature. J Environ Manag 237:408–423. https://doi.org/10.1016/j.jenvman.2019.02.100

Priyadarshanee M, Das S (2021) Biosorption and removal of toxic heavy metals by metal tolerating bacteria for bioremediation of metal contamination: a comprehensive review. J Environ Chem Eng 9(1):104686. https://doi.org/10.1016/j.jece.2020.104686

Rajapaksha AU, Chen SS, Tsang DC, Zhang M, Vithanage M, Mandal S et al (2016) Engineered/designer biochar for contaminant removal/immobilization from soil and water: potential and implication of biochar modification. Chemosphere 148:276–291. https://doi.org/10.1016/j.chemosphere.2016.01.043

Rathi BS, Kumar PS, Show PL (2021) A review on effective removal of emerging contaminants from aquatic systems: current trends and scope for further research. J Hazard Mater 409:124413. https://doi.org/10.1016/j.jhazmat.2020.124413

Richardson SD, Kimura SY (2019) Water analysis: emerging contaminants and current issues. Anal Chem 92(1):473–505. https://doi.org/10.1021/acs.analchem.9b05269

Rodriguez-Narvaez OM, Peralta-Hernandez JM, Goonetilleke A, Bandala ER (2017) Treatment technologies for emerging contaminants in water: a review. Chem Eng J 323:361–380. https://doi.org/10.1016/j.cej.2017.04.106

Rosal R, Rodríguez A, Perdigón-Melón JA, Petre A, García-Calvo E, Gómez MJ et al (2010) Occurrence of emerging pollutants in urban wastewater and their removal through biological treatment followed by ozonation. Water Res 44(2):578–588. https://doi.org/10.1016/j.watres.2009.07.004

Roy N, Alex SA, Chandrasekaran N, Mukherjee A, Kannabiran K (2021) A comprehensive update on antibiotics as an emerging water pollutant and their removal using nano-structured photocatalysts. J Environ Chem Eng 9(2):104796. https://doi.org/10.1016/j.jece.2020.104796

Saleh TA (2020a) Trends in the sample preparation and analysis of nanomaterials as environmental contaminants. Trends EnvironAnal Chem 28:e00101. https://doi.org/10.1016/j.teac.2020.e00101

Saleh TA (2020b) Nanomaterials: classification, properties, and environmental toxicities. Environ Technol Innov 20:101067

Saravanan A, Kumar PS, Jeevanantham S, Karishma S, Tajsabreen B, Yaashikaa PR, Reshma B (2021) Effective water/wastewater treatment methodologies for toxic pollutants removal: processes and applications towards sustainable development. Chemosphere 280:130595. https://doi.org/10.1016/j.chemosphere.2021.130595

Sathishkumar P, Mythili A, Hadibarata T, Jayakumar R, Kanthimathi MS, Palvannan T et al (2014) Laccase mediated diclofenac transformation and cytotoxicity assessment on mouse fibroblast 3T3-L1 preadipocytes. RSC Adv 4(23):11689–11697. https://doi.org/10.1039/C3RA46014B

Shahid MK, Kashif A, Fuwad A, Choi Y (2021) Current advances in treatment technologies for removal of emerging contaminants from water–a critical review. Coord Chem Rev 442:213993. https://doi.org/10.1016/j.ccr.2021.213993

Sharma BM, Bečanová J, Scheringer M, Sharma A, Bharat GK, Whitehead PG et al (2019) Health and ecological risk assessment of emerging contaminants (pharmaceuticals, personal care products, and artificial sweeteners) in surface and groundwater (drinking water) in the Ganges River basin, India. Sci Total Environ 646:1459–1467. https://doi.org/10.1016/j.scitotenv.2018.07.235

Shawaqfeh AT, Al Momani FA (2010) Photocatalytic treatment of water soluble pesticide by advanced oxidation technologies using UV light and solar energy. Sol Energy 84(7):1157–1165. https://doi.org/10.1016/j.solener.2010.03.020

Shu R, Li R, Lin B, Luo B, Tian Z (2020) High dispersed Ru/SiO2-ZrO2 catalyst prepared by polyol reduction method and its catalytic applications in the hydrodeoxygenation of phenolic compounds and pyrolysis lignin-oil. Fuel 265:116962. https://doi.org/10.1016/j.fuel.2019.116962

Sivaranjanee R, Kumar PS (2021) A review on remedial measures for effective separation of emerging contaminants from wastewater. Environ Technol Innov 23:101741. https://doi.org/10.1016/j.eti.2021.101741

Snow DD, Cassada DA, Bartelt-Hunt SL, Li X, D'Alessio M, Levine R et al (2015) Detection, occurrence and fate of emerging contaminants in agricultural environments. Water Environ Res 87(10):868–1937. https://doi.org/10.2175/106143015X14338845155101

Souza RR, Gonçalves IM, Rodrigues RO, Minas G, Miranda JM, Moreira AL et al (2022) Recent advances on the thermal properties and applications of nanofluids: from nanomedicine to renewable energies. Appl Therm Eng 201:117725. https://doi.org/10.1016/j.applthermaleng.2021.117725

Stapleton MJ, Hai FI (2023) Microplastics as an emerging contaminant of concern to our environment: a brief overview of the sources and implications. Bioengineered 14(1):2244754. https://doi.org/10.1080/21655979.2023.2244754

Strady E, Kieu-Le TC, Tran QV, Thuong QT (2021) Microplastic in atmospheric fallouts of a developing southeast Asian megacity under tropical climate. Chemosphere 272:129874. https://doi.org/10.1016/j.chemosphere.2021.129874

Su C, Cui Y, Liu D, Zhang H, Baninla Y (2020) Endocrine disrupting compounds, pharmaceuticals and personal care products in the aquatic environment of China: which chemicals are the prioritized ones? Sci Total Environ 720:137652. https://doi.org/10.1016/j.scitotenv.2020.137652

Sumudumali RGI, Jayawardana JMCK (2021) A review of biological monitoring of aquatic ecosystems approaches: with special reference to macroinvertebrates and pesticide pollution. Environ Manag 67(2):263–276. https://doi.org/10.1007/s00267-020-01423-0

Sun Q, Wada T, Liao CS (2024) Recent advances in bioremediation of emerging contaminants and endocrine disruptors. Front Microbiol 15:1383770. https://doi.org/10.3389/fmicb.2024.1383770

Sutherland DL, Ralph PJ (2019) Microalgal bioremediation of emerging contaminants-opportunities and challenges. Water Res 164:114921. https://doi.org/10.1016/j.watres.2019.114921

Tahira M, Saima M, Rahmat A, Abdul N, Afsar K (2022) Technologies for removal of emerging contaminants from wastewater. In: Muharrem I, Olcay Kaplan I (eds) Wastewater treatment. IntechOpen, Rijeka

Tang Y, Liu Y, Chen Y, Zhang W, Zhao J, He S et al (2021) A review: research progress on microplastic pollutants in aquatic environments. Sci Total Environ 766:142572. https://doi.org/10.1016/j.scitotenv.2020.142572

Tran NH, Reinhard M, Khan E, Chen H, Nguyen VT, Li Y et al (2019) Emerging contaminants in wastewater, stormwater runoff, and surface water: application as chemical markers for diffuse sources. Sci Total Environ 676:252–267. https://doi.org/10.1016/j.scitotenv.2019.04.160

Vargas-Berrones K, Bernal-Jácome L, de León-Martínez LD, Flores-Ramírez R (2020) Emerging pollutants (EPs) in Latin América: a critical review of under-studied EPs, case of study-Nonylphenol. Sci Total Environ 726:138493. https://doi.org/10.1016/j.scitotenv.2020.138493

Vasilachi IC, Asiminicesei DM, Fertu DI, Gavrilescu M (2021) Occurrence and fate of emerging pollutants in water environment and options for their removal. Water 13(2):181. https://doi.org/10.3390/w13020181

Verma M, Singh P, Dhanorkar M (2024) Remediation of emerging pollutants using biochar derived from aquatic biomass for sustainable waste and pollution management: a review. J Chem Technol Biotechnol 99(2):330–342. https://doi.org/10.1002/jctb.7548

Wang J, Zhuan R (2020) Degradation of antibiotics by advanced oxidation processes: an overview. Sci Total Environ 701:135023. https://doi.org/10.1016/j.scitotenv.2019.135023

Wiesner MR, Lowry GV, Alvarez P, Dionysiou D, Biswas P (2006) Assessing the risks of manufactured nanomaterials

Wu J, Zhang L, Yang Z (2010) A review on the analysis of emerging contaminants in aquatic environment. Crit Rev Anal Chem 40(4):234–245. https://doi.org/10.1080/10408347.2010.515467

Zhang S, Wang J, Yan P, Hao X, Xu B, Wang W, Aurangzeib M (2021) Non-biodegradable microplastics in soils: a brief review and challenge. J Hazard Mater 409:124525. https://doi.org/10.1016/j.jhazmat.2020.124525

Zhao S, Liu R, Lv S, Zhang B, Wang J, Shao Z (2024) Polystyrene-degrading bacteria in the gut microbiome of marine benthic polychaetes support enhanced digestion of plastic fragments. Commun Earth Environ 5(1):162. https://doi.org/10.1038/s43247-024-01318-6

Zonja B, Aceña J, Jelic A, Petrovic M, Pérez Solsona S, Barceló D (2014) Transformation products of emerging contaminants: analytical challenges and future needs. In: Transformation products of emerging contaminants in the environment, pp 303–324

Chapter 4
Remediation of Oil Spill: A Menace

Abstract The occurrence of oil spills has remained a major threat of water pollution in the world for the past few decades. These spills have happened as a result of various causes such as vandalism, human error, and mechanical failure among others. Although it is required by law that oil companies must ensure that they are clean up oil spills, an insignificant number of polluted soils in this area have been reclaimed. This has rendered untold hardship to the residents of the communities in this region, as their main source of income, farming and fishing, are being lost in these unfortunate incidents. Thus, the current study assessed the remediation methods that have been applied in the region and other parts of the world with the intention of studying the most ideal methods that should be implemented in this specific area. Prior to the findings of this study, previous research relied mostly on individual techniques, and every technique was known to have its weaknesses. The questions remain whether the integrated understanding of the policy governing the oil industry and the multidisciplinary approach to the selection of remedies could ameliorate the situation.

Keywords Oil spill · Remediation · Environment pollution · Water pollution · Remediation techniques

4.1 Introduction

Human activities contaminate the oceans with oil through the runoff from land, accidents involving vessels, periodic releases from tankers, and discharges from bilges. Oil spills are catastrophic events that have detrimental effects on the environment, affecting not only humans but also various forms of plant and animal life, including avian, aquatic, and terrestrial species. An oil spill refers to the unintentional or intentional discharge of oil into the environment, including both land and water. If the volume of discharged oil exceeds 100,000 gallons, it is classified as a major spill category (Samudro and Mangkoedihardjo 2012; Walther III 2014).

© The Author(s), under exclusive license to Springer Nature
Switzerland AG 2024
N. Saxena et al., *Water Pollution and Remediation*, SpringerBriefs in Water
Science and Technology, https://doi.org/10.1007/978-3-031-76301-4_4

Multiple reasons contribute to the contamination of the maritime environment by oil and its byproducts. The primary and most widespread causes are accidents that occur during the transportation of oil and during oil exploration and production activities. Nevertheless, accidents are responsible for just a small portion of the oil that enters the environment (Ivshina et al. 2015). Marine oil spills have significant negative impacts not only on the marine ecology but also on related activities such as fishing, tourism, marine agriculture, and oil businesses. Industrial activities pose a potential danger of oil leaks, which can cause severe harm to the plant and animal life in the area. According to reports, 30–50% of oil leaks occur due to human errors, whereas 20–40% are caused by either defective machinery or complete equipment failure (Fingas 2010; Michel and Fingas 2016).

4.2   What Is Oil Spillage?

An oil spillage refers to the discharge of unrefined petroleum hydrocarbons into the surrounding environment. An international environmental disaster, usually the result of inadvertent or intentional human activity dumping unrefined petroleum into coastal waterways and land as shown in Fig. 4.1a, b, c. A controversial problem, oil spills can occur during crude oil exploration, oil pipeline leaks, vandalism, unlawful tampering of oil-well heads, oil transfer into boats, and oil transportation in tankers. The oil business gives response to oil spills both upstream and downstream top priority. An oleic slick is formed when oil rapidly disperses across the surface of water, creating a thin coating. As the oil disperses, the layer of oil slick gradually becomes thinner and eventually transforms into a thin layer known as a sheen, which typically exhibits a rainbow-like appearance. This is a common phenomenon, primarily due to the widespread usage of petroleum products by humans, which presents a complex issue now affecting oil-producing areas worldwide.

It is quite common for developed countries to report oil spills and the nature of response actions depending on the severity of the devastation that they bring about. However, in some countries the leakages of the incidents stay unknown and

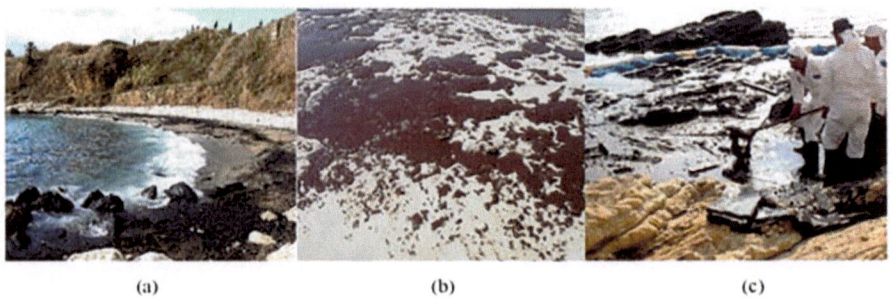

(a)   (b)   (c)

Fig. 4.1 Examples of different forms of oil contamination (Public domain)

sometimes this happens that no proper steps are taken to bring the system back to its initial condition, even though the oil spills are recognized (Anna 2013). Countries maintain the records of oil spill statistics under the supervision of appointed agencies. The International Tanker Owners Pollution Federation Limited (ITOPF) commenced the publication and maintenance of a worldwide database of tanker spills occurrences since this year, 1974. The United States commits to keep the database of spills via US Coast Guard operation, in addition to the Bureau of Safety and Environment (BSEE) who primarily records spills that are related to exploration and production of offshore activities. The National Oil Spill Detection and Response Agency is the body saddled with the responsibility of dealing with oil leakage incidents in Nigeria. Its report states that there are about 5000 leakage sites resulting from the spills in more than 9000 places between the years 2006 and 2015 (Dave and Ghaly 2011). The oil spills in Italy, Canada, South Africa, Angola, France, the Persian Gulf, Uzbekistan, the West Indies, Mexico, and Kuwait ranking among the biggest worldwide have all turned into history. To date, the Kuwait oil spill during the first Gulf War in 1991 is recorded the as the biggest oil spill ever to be recorded. A 240 million gallon of oil was released into the Persian Gulf, forming a slick as wide as the Hawaiian island (Michel and Fingas 2016).

Oil tankers' and offshore disasters have seriously affected the ecology balance globally. Concerns that arise from an oil leak offshore are also a threat to the marine life due to the dangerous impact (Obi et al. 2014). Indeed, once a seafarer spills, the effects are usually much more serious than those on land. Such is the reason that the magnitude of the spills which may extend hundreds of miles over the water zones can within a short time form an ultra-thin layer of oil on the surface. As a result, there have been tremendous efforts towards preventing such major oil spills at the maritime sector. By way of example, in 1989, with 11 million gallons (37,000 metric tonnes) of North Slope crude oil from Exxon Valdez running into Prince William Sound in Alaska, death of many intertidal and subtidal animals resulted and the habitat got affected up to now (Tewari and Sirvaiya 2015).

4.3   Causes of Crude Oil Spills

Oil can get both naturally or influenced by human actions purposefully or unintentionally. The major causes of spills of crude oil and gas by natural events are examples being earthquakes, blag weather and hurricanes. The acts of oil spill caused by humans are technically identified as anthropogenic activities. The possible causes of the accident can be pipeline sabotage, acts, intentional release of oil during a violence and the dumping of ships and vessels in the sea (Chen et al. 2019). The occurrences of fatal oil spills can be varied by the accidental discharge from the ship, technical malfunctions of drilling equipment, the detritus dumping from the processing of the oil tankers, and the hydrocarbon leakage from oil rigs.

4.4   Why Environment Needs to Be Clean Up After Oil Spill?

It impacts the environment heavily with both the oily residue and the allegedly emitted fuel which are the main problems of oil surge. A response to climate change can be divided into two types of effects (physical and chemical) after exposure to alternate climate regimes. The role of the oil spill consequences, in this case, depends on the volume of excreted oil, the chemical and physical characteristics of the polluted oil, the environmental conditions at the time of the spilling and the populations and their inhabitant's strength when facing external issues. Oil spill can harm on living creature as short or long-term effect base on of the exposure, that of the dose might as well increases its toxicity (Asif et al. 2022). By subjecting an organism to an escalated rate of exposure to high levels of pollutants which is comparatively faster than their typical lifespans, the acute effect of a toxic is manifested. Considering the signs and complaints that appear in the fish after 4 days, the following cause of the toxicity acute episode was more likely when the fish were kept out of the contaminants' environment and consequently the signs of the toxicity completely disappeared. Persistent toxicity is what arises in the course of impairment caused by a patient's being exposed to small quantities of pollutants that stick around for very long-time intervals. Seeing the effects is usually done by the consequence that is recorded in the metabolism rate, the growth curves, the reproductive physiology, behaviour and the ability of the species to cope with climatic stresses.

4.5   Crude Oil Effects on Water

Water is the primordial inanimate object and a proven source for sustenance of life. The pollution becomes an influencing factor in life and reduces the quality of the life. The pollution of oils can be capable of changing the behaviour of the very diverse species, by causing abnormalities in their eating, breathing, and reproductive patterns in addition to contaminating and destroying their natural habitats. Polonization is the process of eating a silhouette of varied hydrocarbons by the marine animal which may cause bad taste while eating seafood. Causing seafood to be unfit for human consumption, that taint never vanishes until it is not in the water anymore (Edema 2012). Oil cause significant interference with the cellular functions of marine larvae which leads to the alteration of composition and productivity in the wetlands. This adversely effects the insulation capacity of the fur-bearing mammals such as sea otters. Additionally, the effect is the reduction of the water repellent feature of bird feathering materials. This situation subjects the organisms to a severe environmental condition. During an oil spill, both marine food chains and ecosystem diversity are adversely impacted, and their larval grounds are rendered unliveable, leading to severe toxicity in aquatic species.

4.6 Conventional Methods of Environmental Remediation

Given the frequency of oil spill incidents, it is crucial to possess a comprehensive understanding of their detrimental ecological consequences. This information is essential for developing efficient strategies to restore the environment following an oil leak. Remediation can vary, both ex situ and in situ, by use of various technologies (Hussain 2020). The ex-situ situation combines the cleansing process with the soils and water at a different location, and the in-situ approach treats the same contaminated soils and water directly at the site of pollution.

4.6.1 Technologies for Remediation of Marine Oil Spill

In addition to, removing staining from seas after an oil spill, there are developed other destructive behaviours methods. The application of this technology has helped environmentalists to ward off oil spills which pollution the ocean with time the effects are reduced. To mitigate oil spills in waterways, conventional methods are mainly based on a set of different techniques consisting of physical, chemical, thermal and biological processes respectively (Hussain 2020).

4.6.1.1 Physical Methods

Also called the mechanical methods, these approaches use no chemicals and are popular in several countries. The major purpose of physical techniques such as barriers is to limit the spread of an oil spill.

(a) Booms

Oil booms are devices employed to capture and hold oil spills. They play two main roles as a physical barrier that puts drifting oils inside them and blocks the dispersion of them. In addition, the booms are used to direct oil from environmentally sensitive places, or to consolidate oil so that its viscosity is suitable for the use of skimmers or other cleansing ways (Cai et al. 2021). Such techniques serve to extract oil from the spill site. Most boom designs can be classified into two main categories: battens and fence rails. Oil slick booms are created using fire-resistant metals that can resist the blazing heat that is produced by burning oil. The fire booms can be with a fence or a curtain design. Additional specialized booms types are coastal seal booms, ice booms intended for weak ice conditions, sorbent booms, and tidal seal booms. Fence booms are the buoyant structures that float on the water's surface. The flat section is provided and they are positioned vertically either by built-in or external buoyancy. They have a light weight, are easy to carry, are hygienic and occupy less storage space. Furthermore, these fibres are resilient to abrasion. Conversely, the main shortcomings of the fence booms are the reduced durability in heavy wind and current, the limited capacities for towing, and the reduced effectiveness in rough sea.

(b) Skimmers

Oil skimmers combined with booms are frequently used to extract oil from the ocean surface without changing the water characteristics. The oil is then delivered to storage tanks on the vessel through the transportation system. This method may not change the oil's composition and nature, making it possible to recycle it and reuse it (Đorđević et al. 2022). The effectiveness of skimming depends upon the type and viscosity of spilled oil and the existing weather condition. Skimming devices usually work not too bad in quiet water but can get clogged from floating trash. These skimmers are divided into three types of oil skimmers: the oleophilic, suction, and weir skimmers. The main operation principles of skimmers such as special gravity effect, surface tension, and a mobile medium help to effortlessly deal with oil contamination on the water surface. They may be fully autonomous, operated from land or remotely, water-based.

Hydrophobic skimmers employ materials which have a tendency to cling to oil. The oil that is collected on the surface is then scraped or squeezed into a recovery tank as wastewater. The oleophilic materials can be of different shapes like disc, drum, belt, brush, rope-mop. Suction skimmers are routinely applied for oil recovery in different contexts like beaches, a limited scope of the surface, and low-lying lands. There is a special machine which is called a suction skimmer—it is a vacuum pump with a wide floating head and delivers oil into storage tanks. The principle of gravity is used in weir skimmers for water surface oil gathering (skimming). The weir works as a barrier containing the oil into the washbasin's centre. From there, it is pumped to a storage tank and used for wastewater recycling.

(c) Adsorbent Materials

Adsorbent materials promote the change of the liquid into the semisolid state while aid after skimming the process to purify the remained oil. These adsorbents may be based on natural organic, natural inorganic, and synthetic materials. Natural organic sorbents can utilize peat, hay, vegetable fibres, feathers, kapok, sawdust, milkweed, and straw, chitosan as examples (Ahmed et al. 2020). Researchers have leveraged these readily available and economical materials to reach an absorption peak. They can soak up the grease from anywhere between 3 to 15 times their own weight. One of the big deficiencies of their utilisation is the problem of collecting and disposing of adsorbents after they have been expanded on the surface of the contaminated waters. Illustrations of natural inorganic sorbents include among others vermiculite, clay, wool, glass, sand, vermiculate, and volcanic ash. Furthermore, they can be readily available and possess the capability to absorb about four to twenty times their own volume. The most common sorbent compounds in use are artificial absorbents. The members of the list consist of polyethylene, polypropylene, polyurethane foam, polyester foam, polystyrene and nylon fibbers which are the same as plastics. Because of their hydrophobicity and oleophilicity, they have the capacity to absorb oil up to 70–100 times their own weight. In addition, they are not biodegradable.

4.6.1.2 Thermal Method

The process involves the burning of oil in this fire-resistant equipment like boom fire or igniters. Having in place a burning strategy immediately after a spill, leads to the most desirable recovery level. An efficient and cost-effective way to tackle oil spills in calm and open waters, especially suited to refined oil products, which ignite quickly without posing a danger to animals in the marine ecosystem. On the first step of this thermal method, the nearby floating vessel should be checked for being any ship, speedboat or the oil tanker. The major drawbacks in the practice of the 'controlled' approach are the threat of secondary fires, the destruction of adjacent plant and animal life caused by burning, the harm that aquatic plants and animals face even many after the burn, and the human health risk that comes from burning gas. Water temperature, speed, wave amplitude, wind direction, slick thickness, oil type, and levels of weathering and emulsification are important factors in river oil spill thermal remediation (Tewari and Sirvaiya 2015).

4.6.1.3 Chemical Methods

The chemical technologies used in the marine ecosystem not only prevent the oil spill from further spreading to the shoreline, but also safe-guards the sensitive marine populations and habitats. The clean-up processes involve most of the impressive chemical techniques which leads to one of the best remediations for both onshore and offshore environments (Tewari and Sirvaiya 2015). Such techniques can be successfully employed in combination with dissolvers and gelling agents. Both of these substances change the oil's characteristics either physic or chemically.

(a) Dispersants

Surfactants are interface-active materials that are present in the dispenser. Dispersants‚Äô primary purpose is to disperse oil slicks into many tiny droplets that then sinks into the lower parts of the water column, quickly mix with the surrounding water and become easier to break down (Merlin et al. 2021). Dispersants containing highly concentrated chemicals include SlickgoneNS, Neos AB3000, Corexit 9500, Corexit 8667, Corexit 9600, SPC 1000 T, Finasol OSR 52, Nokomis 3-AA and Nokomis 3-F4, Saf-Ron Gold, Z (Varasteh et al. 2024). The dispersants are delivered by spraying them into the water through the use of ships or aircraft. This method of surface application is especially effective in turbulent sea conditions as it produces a good mix between the chemical and water. Each and every dispersion molecule has the oleophilic nature (that is attraction to oil) as well as the hydrophilic nature (which means attraction to water). As the solvent spreads the dispersants at the oil and water interface, the molecules are reorganized in such that the oil -loving segment of the molecule is presented to the oil and the water loving segment is presented to the water. The wetting process downregulates the interfacial tension between oil and water, allowing wave energy to be added to the separation process of oil droplets from the oil slick. Dispersants act to accelerate the removal of contaminants from water, crudely eliminate the formation of oil-water mixtures, greatly reduce the likelihood of oil sticking on surfaces (such as marine animals including seabirds), and expedite the normal occurrence of biodegradation. However, the increasingly combustible properties of the dispersants represent a prominent danger for both humans' health and the marine environment.

(b) Solidifiers

Solidifiers are a hydrophobic material which is made less hydrophobic by the Van der Waals forces which spread out the material and make it harder. Those kinds of microorganisms turn the oil into a solid rubber that is not having density and can be encountered physically with the environment, such as skimming. They are hydrophobic (oliophilic) materials and may remain in the form of dry particles or semi-solid solids. There exist three distinct categories of solidifiers, each with distinct traits and properties: such as polymeric sorbents, cross-linking agents, and polymers co-immobilized with cross-linking agents. To mention such, as solids, are oil flex, Norsores, oil bond, Molet wax, Elastole, Gelco 200, CI agent, rubberizer, Jett gel, and Smart bond HO used by (Fingas and Fieldhouse 2011).

4.6.1.4 Biological Method

Bioremediation, or the use of biological technique, is the avenue through which microorganisms completely breakdown and metabolise chemical pollutants, so as to restore the environment (Bala et al. 2022). The underlying principle of bioremediation is the repair of the environment at sustainable and environmental-friendly cost. Bioremediation harnesses either the natural ability of microorganisms to metabolise a wide array of pollutants, or it stimulates them through bio-stimulation or augments them through bioaugmentation. Bio-stimulation and bioaugmentation are two approaches in bioremediation.

Bio-stimulation refers to nutrient supplementation to promote growth of microbes, while bioaugmentation refers to introduction of microbes to a population with inherent capacity for degradation of oil. There are only a few bacteria that can metabolise petroleum hydrocarbons. Organic molecules will become incorporated into bacterial cell biomass and, at the same time, carbon dioxide, water and heat will be produced. Each family of microorganisms uses a distinct mechanism for biomineralization of hydrocarbons. Microorganisms that can degrade hydrocarbons are found throughout the area affected by a marine oil spill. Paraffinic and aromatic hydrocarbons can be degraded by a variety of microorganisms, but they degrade at different rates. Biodegradation of oil in the marine environment is mainly dependent on availability of nutrients, concentration of oil, duration of time and extent of spontaneous biodegradation.

4.7   Advantages and Disadvantages of Conventional Methods

The mechanical method of removing oil spills is unsuitable for rough sea waves, high velocity of wind and high amplitude of water wave, even though it is low cost and easy to collect. The major problems of employing natural inorganic adsorbent materials include that they cannot be compatible with the surface of water, they need lots of labour input and high rate of both water and oil adsorption, so they will sink; several natural inorganic adsorbents, for example, clay and vermiculite are loose materials, which is difficult to use in the windy situations, and health hazard will be a concern if they are inhaled. Chemical dispersants have great efficiency, but their toxic properties and long-term environmental damages are major worries. The natural overall reasons for the publicity of biological ideas are soil sustainability and soil can be recovered to the original state (Zamora-Ledezma et al. 2021).

4.8   The Use of Biological Techniques in Crude Oil Spill Cleanup

Biological oil spill cleaning entails employing flora and fauna to conduct the remediation of contaminated areas, specifically addressing the cleanup of crude oil spills. Bioremediation and phytoremediation, which are biological treatments, have been used to clean up places contaminated with oil. Bioremediation, as defined by Frick et al. (1999), refers to the use of biological processes to detoxify contaminated areas. Bioremediation employs microorganisms and their products to eradicate pollutants from the environment (Leung 2004), while phytoremediation improves this process in the presence of plants. As a result, the process of addressing and resolving crude oil pollution has achieved different degrees of effectiveness worldwide.

4.8.1   Bioremediation

Modern bioremediation refers to the utilisation of microorganisms to cleanse contaminated environments, and George Robinson is partially credited with its development. George Robinson effectively employed microorganisms to mitigate an oil leak along the Santa Barbara, California coastline in the late 1960s. Subsequently, particularly from the 1980s onwards, there has been significant focus on oil spills and the bioremediation of hazardous waste (Shannon and Unterman 1993).

Mohamed (2016) states that the bioremediation technology is widely recognised worldwide as an in-situ treatment and is an efficient strategy to manage and recover the contaminated environment. Bioremediation utilises microorganisms and their byproducts to eliminate pollutants from the environment (Leung 2004). These microorganisms have the ability to eradicate, convert, confine, or diminish the strength of pollutants found in the environment. Shannon and Unterman (1993) state that the process specifically focuses on chemicals via means of transformation, mineralization, or modification. According to Yakubu (2007), the intentional application of bioremediation to reduce toxic waste materials is a relatively new concept, especially when contrasted to its usage in wastewater treatment. Bioremediation is a redox reaction that generates energy through respiration and other biological processes necessary for the growth and upkeep of microbial cells, where these reactions occur.

4.9 Conclusion

There should be a stronger focus on remediation of oil spillage in order to minimise the resulting catastrophe caused by oil spills. It is important to select repair methodologies with caution in order to guarantee optimal efficiency and efficacy. Biological remediation is particularly advantageous because to its proven sustainability and cost-effectiveness, notwithstanding the effectiveness of other processes such as physical, chemical, thermal, and biological methods in reaching this goal.

References

Ahmed MJ, Hameed BH, Hummadi EH (2020) Review on recent progress in chitosan/chitin-carbonaceous material composites for the adsorption of water pollutants. Carbohydr Polym 247:116690

Anna MA (2013) Evaluation of the methods for the oil spill response in the off-shore arctic region (Bachelor's thesis). Bachelor of engineering degree programme in environmental engineering. Helsinki Metropolia University of Applied Sciences, p 52

Asif Z, Chen Z, An C, Dong J (2022) Environmental impacts and challenges associated with oil spills on shorelines. J. Mar. Sci. Eng. 10(6):762

Bala S, Garg D, Thirumalesh BV, Sharma M, Sridhar K, Inbaraj BS, Tripathi M (2022) Recent strategies for bioremediation of emerging pollutants: a review for a green and sustainable environment. Toxics 10(8):484

Cai J, De Silva DG, Slechten A (2021) Effects of oil booms on the local environment. Energy Econ 101:105365

Chen J, Zhang W, Wan Z, Li S, Huang T, Fei Y (2019) Oil spills from global tankers: status review and future governance. J Clean Prod 227:20–32

Dave DAEG, Ghaly AE (2011) Remediation technologies for marine oil spills: a critical review and comparative analysis. Am J Environ Sci 7(5):423

Đorđević M, Šabalja Đ, Mohović Đ, Brčić D (2022) Optimisation methodology for skimmer device selection for removal of the marine oil pollution. J. Mar. Sci. Eng 10(7):925

Edema N (2012) Effects of crude oil contaminated water on the environment. In: Crude oil emulsions–composition stability and characterization, pp 169–180

Fingas M, Fieldhouse B (2011) Surface-washing agents or beach cleaners. In: Oil spill science and technology. Gulf Professional Publishing, pp 683–711

Fingas M (ed) (2010) Oil spill science and technology. Gulf professional publishing

Frick CM, Germida JJ, Farrell RE (1999, December) Assessment of phytoremediation as an in-situ technique for cleaning oil-contaminated sites. In technical seminar on chemical spills (pp. 105a-124a). Environment Canada, 1998

Hussain CM (ed) (2020) The handbook of environmental remediation: classic and modern techniques. Royal Society of Chemistry

Ivshina IB, Kuyukina MS, Krivoruchko AV, Elkin AA, Makarov SO, Cunningham CJ et al (2015) Oil spill problems and sustainable response strategies through new technologies. Environ Sci: Processes Impacts 17(7):1201–1219. https://doi.org/10.1039/C5EM00070J

Leung M (2004) Bioremediation: techniques for cleaning up a mess. BioTeach J 2:18–22

Merlin F, Zhu Z, Yang M, Chen B, Lee K, Boufadel MC et al (2021) Dispersants as marine oil spill treating agents: a review on mesoscale tests and field trials. Environ Syst Res 10:1–19

Michel J, Fingas M (2016) Oil spills: causes, consequences, prevention, and countermeasures. In: Fossil fuels: current status and future directions, pp 159–201. https://doi.org/10.1142/9789814699983_0007

Mohamed EA (2016) Biophysical removal of some toxic heavy metals by Aeromonas strains. J Pure Appl Microbiol 10(1):311–316

Obi EO, Kamgba FA, Obi DA (2014) Techniques of oil spill response in the sea. IOSR J Appl Phys 6:36–41

Samudro G, Mangkoedihardjo S (2012) Assessment framework for remediation technologies of oil polluted environment. Int J Acad Res 4(2):36–39

Shannon MJ, Unterman R (1993) BIOR. EMEDIATION: distinguishing fact from fiction. Annu Rev Microbial 715:38

Tewari S, Sirvaiya A (2015) Oil spill remediation and its regulation. Int J Eng Res Gen Sci 1(6):1–7

Varasteh T, Lima MS, Silva TA, da Cruz MLR, Ahmadi RA, Atella GC et al (2024) The dispersant Corexit 9500 and (dispersed) oil are lethal to coral endosymbionts. Mar Pollut Bull 203:116491

Walther III HR (2014) Clean up techniques used for coastal oil spills: an analysis of spills occurring in Santa Barbara, California, Prince William sound, Alaska, the sea of Japan, and the Gulf coast

Yakubu MB (2007) Biological approach to oil spills remediation in the soil. Afr J Biotechnol 6(24):2735–2739

Zamora-Ledezma C, Negrete-Bolagay D, Figueroa F, Zamora-Ledezma E, Ni M, Alexis F, Guerrero VH (2021) Heavy metal water pollution: a fresh look about hazards, novel and conventional remediation methods. Environ Technol Innov 22:101504

Chapter 5
Green Remediation Technologies

Abstract During repair operations, green remediation approaches involve assessing the undesirable ecological effects of water treatment and developing alternative alternatives to reduce harm and enhance good impacts on the environment. This chapter will comprehensively analyse the research on photocatalysis for the removal of water pollution. The topic will cover progress in photocatalysts, process factors, efficacy levels, and widespread deployment. The authors explain many methods of using phytoremediation tactics to reduce, eliminate, and address the detrimental effects of contaminants discharged through the ecosystem. Utilising plants to reduce soil contaminants is an economically efficient method that mitigates the risks to both people's well-being and the environment resulting from polluted water bodies. We provide an in-depth analysis of the basic principles of microbial fuel cells (MFC) and their dual capability to remove pollutants and generate electricity from wastewater. Investment viability requires further investigation, especially in economic research, to evaluate the potential of these technologies.

Keywords Green remediation · Photocatalysis · Phytoremediation · Microbial fuel cell · Sustainable · Eco-friendly

5.1 Introduction

Green remediation techniques involve considering the adverse ecological consequences of water treatment and finding alternative solutions to minimise harm and maximise positive impacts on the environment during remediation operations. Through collaboration, these green remediation techniques aim to enhance the production of sustainable and green energy, ensuring both cost-effectiveness and environmental friendliness. When it comes to addressing water pollution, the utilisation of photocatalysis, phytoremediation, and different fuel cell technologies will play a vital role.

Photocatalytic degradation of pollutants has gained significant attention as an effective approach to remediating contaminated water (Malini and Gandhimathi 2024; Sathishkumar 2024; Loeb et al. 2018). As we know, water pollution is causing

N. Saxena et al., *Water Pollution and Remediation*, SpringerBriefs in Water
Science and Technology, https://doi.org/10.1007/978-3-031-76301-4_5

severe harm to both ecosystems and human well-being. The utilisation of advanced oxidation processes (AOPs) enables efficient decomposition of organics into complete mineralization (i.e., CO_2 and H_2O) of waste by roducing reactive oxygen species (ROS) (Islam et al. 2023). The hydroxyl ($\bullet OH^-$) and superoxide ($\bullet O_2^-$) radicals are the most well-known ROS. O_3 and H_2O_2 are examples of chemical oxidants used in traditional AOPs. Often, a vitality source like ultraviolet (UV) treatment is used to generate ROS for the purpose of oxidising and eliminating water contaminants (Loeb et al. 2018). Photocatalytic water purification is generally acknowledged to be more effective than homogeneous-phase advanced AOPs. Industrial applications widely use titanium dioxide (TiO_2) as their preferred semiconductor due to its cost-effectiveness, strong physical durability, and low level of toxicity (Hoffmann et al. 1995). It may be stimulated by low-energy ultraviolet radiation (UV-A), which makes it appropriate for use in solar energy applications.

Similarly, phytoremediation is an additional eco-friendly type of cleanup. Phytoremediation is an economically efficient and ecologically harmless method of remediation that has the capacity to function as a promising preventive measure against water contamination, particularly heavy metals (Nedjimi 2021). Understanding the mechanisms involved in the buildup of heavy metals and the ability of plants to tolerate them is essential for improving the effectiveness of phytoremediation. This topic examines the mechanisms by which plants absorb, transport, and neutralise harmful substances, namely heavy metals.

Microbial fuel cells (MFCs) enable the production of environmentally friendly energy and the removal of contaminants from wastewater simultaneously (Nawaz et al. 2022). The rising popularity of MFCs can be attributed to their dual capability of simultaneously producing electricity and purifying wastewater. The recent research on MFCs used for both wastewater treatment and energy generation is analysed.

This chapter will thoroughly examine the studies on photocatalysis for the purpose of removing water pollution. The discussion will encompass advancements in photocatalysts, variables in the process, the level of effectiveness, and the extensive implementation. The authors describe the many ways in which phytoremediation occurs. The removal of various pollutants through phytoremediation is discussed. Examination of the fundamental MFC principle and its ability to eliminate contaminants and produce electricity from wastewater. Investment viability necessitates future research, particularly in the realm of economic research, to assess the potential of these technologies.

5.2   Photocatalysis

Visualise a static depiction of the sun positioned towards the bottom of the horizon. Is it ascending or descending? Predicting the future course of it can only be done by analysing consecutive time-lapse photography. The progression of a novel technology follows a comparable trajectory, albeit with an indeterminate duration, ascending from its initial inception to attain the pinnacle of popularity before descending towards obscurity (Loeb et al. 2018). Photocatalytic AOPs for water treatment have undergone significant technological advancements during the past 35 years. Traditional AOPs use chemical oxidants like O_3 and H_2O_2 and energy sources like UV irradiation to create ROS that break down pollutants in water through oxidation. Photocatalysis by heterogeneous semiconductors helps with contemporary oxidation in a unique way: when a catalyst is used to absorb light whose energy is higher than or equal to the material's band gap, it creates a pair of conduction band electrons $\left(e_{CB}^-\right)$ and valence band holes $\left(h_{VB}^+\right)$. Hydroxyl radicals (•OH) are considered the primary oxidizing agent for water treatment. Photocatalysis, which originated from the observation of photoelectrochemical water splitting on a TiO_2 electrode (Fujishima and Honda 1972), was promptly acknowledged for its capacity in water treatment (Malini and Gandhimathi 2024; Sathishkumar 2024; Okamoto et al. 1985; Hoffmann et al. 1995).

5.2.1 Mechanism of Pollutants Degradation

Photocatalysis is an AOP that has demonstrated the potential for practical application in real-world scenarios. The procedure does not include the use of any chemicals or gas emissions that could potentially lead to additional contamination. Moreover, the photocatalytic system possesses the ability to oxidise and transform challenging pollutants into more manageable forms (Sathishkumar 2024; Khan et al. 2015). Fig. 5.1 demonstrates the basic process of photocatalytic degradation (Islam et al. 2023). The photocatalyst initially absorbs light, leading to the creation of CB electrons and valence band holes. This is the initial phase of the procedure. The second stage of the photocatalytic procedure entails the transportation or diffusion of environmental pollutants to the photocatalyst's surface. In the third stage, a process known as adsorption attracts pollutants to the photocatalyst's surface. During the fourth stage, the photocatalyst's surface facilitates the oxidation and reduction (redox) reactions to minimise the adverse effects of pollutants. The pollutant molecules are absorbed by the VB and CB. The fifth phase then extracts (desorbed) the products from the photocatalyst's surface. The final stage (Elsalamony 2016) involves the transportation or dispersion of substances into the liquid solution. According to Anwer et al. (2019), we can understand the mechanisms involved in the degradation of pollutants through photocatalysis.

$$Photocatalyst + hv \rightarrow h_{VB}^+ + e_{CB}^- \tag{5.1}$$

$$h_{VB}^+ + e_{CB}^- \rightarrow Energy\left(heat\right) \tag{5.2}$$

$$H_2O + h_{VB}^+ \rightarrow OH^{\bullet}\left(Hydroxyl\ radical\right) + H^+ \tag{5.3}$$

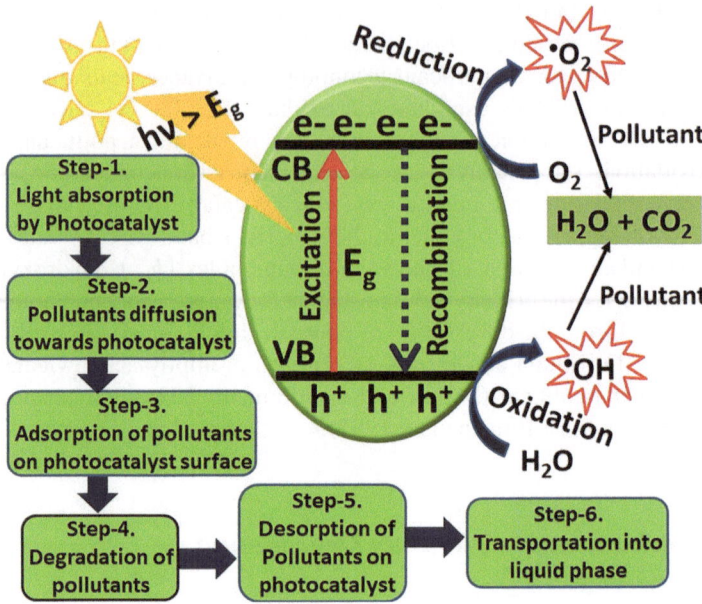

Fig. 5.1 The process of degrading pollutants by photocatalysis. (Reproduced from ref (Islam et al. 2023), Copyright (2023), with permission from John Wiley and Sons)

$$O_2 + e_{CB}^- \rightarrow O_2^{\cdot} (Superoxide\ radical \tag{5.4}$$

$$OH^{\cdot} + Pollutant \rightarrow Intermediates \rightarrow H_2O + CO_2 \tag{5.5}$$

$$O_2^{\cdot} + Pollutant \rightarrow Intermediates \rightarrow H_2O + CO_2 \tag{5.6}$$

5.2.2   Fundamental Barriers to Research-to-Practice

Photocatalysis, a treatment method for surface and groundwater contaminants, has shown limited adoption in water treatment practices due to challenges in economic viability. Research on photocatalytic AOPs has been limited, often exaggerating its advantages and downplaying its drawbacks. Academic literature often focuses too much on material design and mechanistic evaluations, resulting in a gap between successful academic research and limited commercial implementation. This has led to increased scepticism and doubt about the extensive implementation of photocatalysis for water purification method in the next 20 years (Sathishkumar 2024; Loeb et al. 2018).

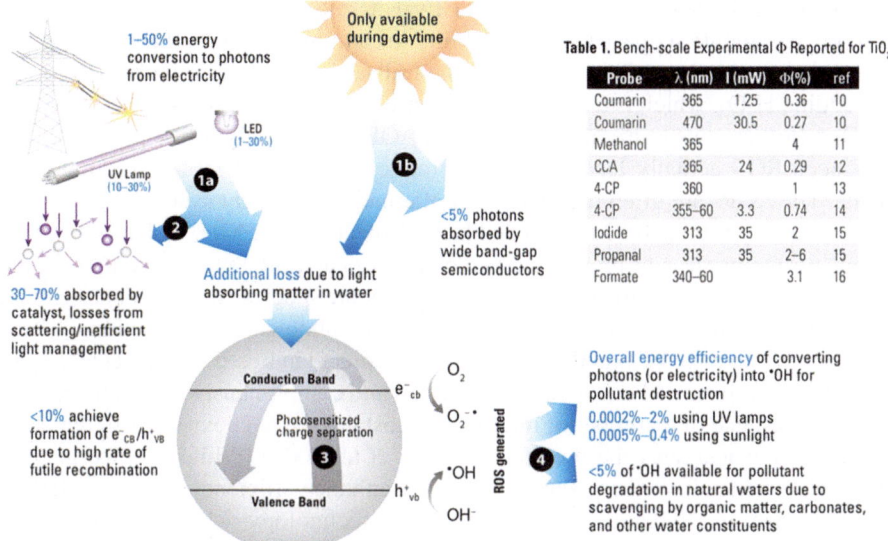

Table 1. Bench-scale Experimental Φ Reported for TiO$_2$

Probe	λ (nm)	I (mW)	Φ(%)	ref
Coumarin	365	1.25	0.36	10
Coumarin	470	30.5	0.27	10
Methanol	365		4	11
CCA	365	24	0.29	12
4-CP	360		1	13
4-CP	355–60	3.3	0.74	14
Iodide	313	35	2	15
Propanal	313	35	2–6	15
Formate	340–60		3.1	16

Fig. 5.2 Outlining the standard procedures for TiO$_2$ photocatalytic water treatment and their low-energy conversion yields. Two main methods exist for producing photons: electrical conversion and direct solar energy harvesting. (1a) In contrast to low-pressure mercury lamps, blue/UVA LEDs with wall plug efficiencies exceeding 30% are being researched (Khan et al. 2008). (1b) Sun energy eliminates electrical conversion losses; however, TiO$_2$'s broad band gap mismatches the sun spectrum, resulting in <5% photon absorption (Loeb et al. 2016). (2) Ineffective light control and inherent scattering can cause 30–70% of the photons produced in a reactor to be absorbed catalytically (Grčić and Puma 2017). (3) Energy from light that the catalyst absorbs creates e_{CB}^- and h_{VB}^+ and leads to ROS generation through surface redox reactions of catalysts. More than 90% of photogenerated e_{CB}^- / h_{VB}^+ pairs undergo recombination quickly (sub μs) (Hoffmann et al. 1995), resulting in fewer than 10% quantum yields (Φ) for most materials. (4) Scavenging significantly lowers the quantity of ROS available for pollutant destruction. (Reproduced from ref (Loeb et al. 2018), Copyright (2018), under Creative Commons Open Access Licence from American Chemical Society)

The low photoconversion efficiency remains a significant obstacle. Figure 5.2 provides a full explanation of how each phase of the photoconversion process results in a decrease in efficiency. The quantum yields (Φ) estimated in the literature are always very low and can vary a lot under different experimental circumstances and with different photocatalysts. Even when employing an imaging base and single-color light close to the band gap (Fig. 5.2), they remain negligible, if not non-existent. The poor efficiency of ˙OH generation is regarded as the most significant limitation for the purification of water via photocatalysis when compared to other AOPs. Using UV/H$_2$O$_2$ AOPs as an illustration, the process of breaking down H$_2$O$_2$ through photolysis in water without organic compounds at a wavelength of 254 nm has been reported to have a quantum yield (Φ) of 50% and a nearly 100% yield in the formation of ˙OH radicals (Mierzwa et al. 2018). This high efficiency makes it extremely difficult for photocatalysis to rival it in terms of energy efficacy. It is important to mention that TiO_2 has the ability to capture photons in the UV-A region, which is less energetic. This allows for the utilisation of sunlight without the need for adding H_2O_2, which is a significant expense in traditional AOPs.

One further constraint that applies to all AOPs is that only a tiny portion of the produced ROS actually leads to the eventual elimination of the desired contaminants. The AOPs are often preferred over alternative treatment systems due to their high reactivity (about 10^9 M^{-1} s^{-1}) and lack of specificity of ·OH, which is advantageous for the process of breaking down many organic pollutants. Nevertheless, the presence of natural organic matter (NOM), carbonate species, and other background elements can significantly reduce the efficacy of photocatalytic reactions. These substances scavenge ROS and absorb light, as depicted in Fig. 5.2 (Katz et al. 2015). In fact, research carried out at both the trial run and comprehensive levels has shown that the presence of a number of interfering substances that coexist at different concentrations exacerbates these effects (Benotti et al. 2009). While engineering and materials science have the potential to greatly enhance efficiency, it is important to acknowledge the inherent limitations of the technology when applying fresh research findings to industrial settings.

5.2.3   Pursuit of Better Photocatalytic Solutions

The continuous production investigations into the advancements regarding the development of novel photocatalytic materials remain unaffected by the low level of industry implementation. A growing amount of research has concentrated on the creation of improved catalyst substances, driven by advancements in materials science and nanotechnology. The initial endeavours involved alterations to anatase TiO_2. For instance, the chemical reaction of the flame pyrolysis process is a large-scale, cost-effective method that has been shown to add a small amount of rutile phase to P25 TiO_2 powder, which makes it more photocatalytic (Sathishkumar 2024; Hurum et al. 2003). Undoubtedly, TiO_2 remains a focal point in the field of photocatalytic materials science research. This research primarily focuses on addressing two major drawbacks of TiO_2 as a means to enhance effectiveness of catalysts: its reduced ability to absorb light and a fast rate of combination between the two main photogenerated species, e_{CB}^- and h_{VB}^+.

We have successfully achieved significant enhancements to TiO_2. The expansion of light-based energy at shorter wavelengths was initially achieved by including transition metals (such as V, Cr, and Fe) (Choi et al. 2002; Suri et al. 1993) and subsequently through the implementation of non-metallic dopants (such as S, C, N, and F) (Banerjee et al. 2014) to generate oxygen vacancies or localised energy levels at the dopant's low-lying inter-band states. A recent advancement in this method was the use of surface hydrogenation to generate several mid-gap states caused by disorder. This caused the valence band edge to migrate upwards, leading to the formation of black or blue-coloured TiO_2 crystals that absorb light in the infrared range

(Chen et al. 2011). While enhancing the absorption of visible light, an excessive number of mid-gap states might result in increased recombination and a constricted band structure. This, in turn, reduces the redox potential of the e_{CB}^- / h_{VB}^+ pair and affects the specific ROS generated (Zhao et al. 2014). Semiconductors with a lower band gap can form heterojunctions, expanding the absorption range to include visible light. Additionally, the absorption range can be expanded by sensitising with organic chromophores. Furthermore, noble metal nanoparticles that absorb visible light have the ability to introduce electrons into the CB of the photocatalyst (Tsukamoto et al. 2012).

Metal junctions serve as electron-withdrawing centres in synthesis, reducing recombination and promoting stability by way of creation regarding a mending Schottky barrier (Xiong et al. 2011). However, excessive cocatalysts have the potential to restrict the surface area and introduce complexity in the process. Semimetallic carbonaceous materials like graphene (Zhang et al. 2010) can produce similar composite architectures, but their inherent instability in oxidative environments remains a concern. Nanoscale manipulation of TiO_2 structure and porosity can enhance its effectiveness by facilitating charge movement and maximising surface area (Sathishkumar 2024; Loeb et al. 2018). Scientists are currently focused on enhancing the efficiency and reliability of these enhanced TiO_2 compounds.

Water treatment has been extensively explored using alternative semiconductor materials, such as cadmium sulphide (CdS) and tungsten trioxide (WO_3), as photocatalysts for visible light. However, the presence of hazardous elements like cadmium is incompatible with using secure and environmentally friendly methods of water purification. Liu et al. (2015) say that graphitic carbon nitride (g-C_3N_4) is a photocatalyst that can be activated by visible light but is not chemically stable. Perovskite materials, which are made up of both organic and inorganic parts, can be used to change how charges are absorbed and moved, but they are very unstable when they get wet (Wang et al. 2015). Platinized tungsten oxide (Pt/WO_3) has been identified as a potential substitute for TiO_2, generating ·OH via multielectron reduction when exposed to visible light (Kim et al. 2010). $BiPO_4$, a recently discovered alternative, exhibits a maximum VB potential that is more positive than that of TiO_2, enhancing oxidation power, stronger photoactivity, and higher efficacy in mineralization (Pan and Zhu 2015). However, it requires higher-energy UV excitation, which comes at a cost. Innovative and undiscovered solutions are needed to apply these substances, which show promise for use in water purification.

Research in photocatalysis materials science has led to the development of advanced materials, such as TiO_2, but their superiority remains a challenge. The evaluation of these materials, which often overlook practical issues, is crucial in determining their superiority. Despite the growing number of scientifically intriguing materials, they are often impractical due to their fragility, chemical instability, or the presence of costly, scarce, or harmful elements. Improving the evaluation criteria to include factors such as better spectral match, less scattering, cost-effectiveness, long-term stability, durability, and ease of separation would make it easier to use the technology in industry (Sathishkumar 2024; Loeb et al. 2018).

5.2.4   Engineering and Assessment of Photocatalytic Reactor Systems

The emphasis on studies focusing on material science instead of reactor or systems engineering potentially neglects the potential for enhancing efficiency regarding the enhancement of photocatalytic water purification technologies via creative reactor layout. Although there has been much research conducted in this field (Van Gerven et al. 2007), the practical implementation of sparse configurations has mostly been limited to reactor installation, where light is projected into a homogeneous mixture of well-dispersed TiO_2 nanoparticles in order to maximise the absorption of photons and the movement of redox species (Benotti et al. 2009), as depicted in Fig. 5.3a. These systems have the potential to compete with other advanced oxidation processes (AOPs) in specific specialised applications, as mentioned later. However, they are not yet fully optimised and have drawbacks such as low energy efficiency, slow reaction rates, and, in some circumstances, issues with catalyst fouling or

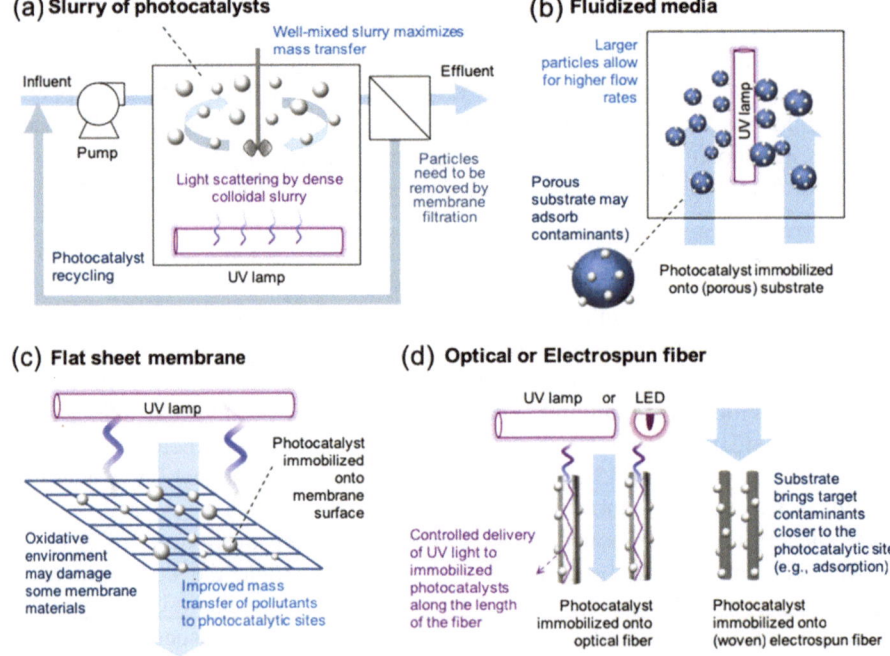

Fig. 5.3 Consider the following reactor designs for treating water using semiconductor photocatalysis: (**a**) A typical slurry reactor setup includes a reactor with TiO2 photocatalysts mixed in a liquid, a quartz low-pressure mercury ultraviolet light, and a filtering membrane system to get the catalysts back. (**b-d**) Newly designed reactor layouts utilise immobilised catalyst particles, eliminating the requirement for subsequent membrane treatment for separation. (Reproduced from ref (Loeb et al. 2018), Copyright (2018), under Creative Commons Open Access Licence from American Chemical Society)

photo aggregation (Benotti et al. 2009; Puma and Brucato 2007). Alternatively, the process of immobilising photocatalysts onto support substrates can be employed, thereby avoiding the requirement for the process of separating substances using ultrafiltration and lowering the shear pressure exerted upon particles of catalyst. However, the effectiveness of immobilised systems can be additionally constrained by factors such as diminished catalyst surface area, decreased lit catalyst surface area per unit volume of water with treatment, and elevated rates of photon scattering. The study has investigated various immobilisation substrates, such as optical fibres, foam porous supports, fluidized media, electrospun fibres, and membranes. These substrates have been specifically designed to enhance pollutant destruction while minimising the removal of reactive oxygen species (ROS). Figure 5.3b-d illustrates the exploration of these substrates. Most of these immobilised reactor concepts are still in the experimental stage or in the initial stages. The recent quick progress in light-emitting diode (LED) technology is expected to stimulate the development of new and innovative reactor designs. LEDs provide several advantages over gas discharge lamps, such as more durable casings, the absence of hazardous substances, their compact size, and their quick getting-ready (Matafonova and Batoev 2018; Chen et al. 2017).

Comparing photocatalytic systems and other AOP technologies is challenging due to the prevailing metric of electrical energy per unit order (EE/O). This metric is used for assessing energy efficacy and emissions output for typical pollutants, allowing for a smooth transition from small-scale testing to larger-scale implementation. For drinking water applications, values ranging from 0.5 to 10 kWhm^{-3} are considered competitive. However, standard TiO$_2$/UV slurry reactors typically have values of 10 kWhm^{-3} or greater. AOPs like UV/H$_2$O and H$_2$O/O$_3$ can achieve values of less than 1 kWhm^{-3}. It is important to consider additional aspects, such as the energy requirements or expenses associated with using consumable chemicals, when comparing systems based on a single figure (Loeb et al. 2018).

5.2.5 Measures to Improve Research Results

It is possible that chances to enhance the efficacy of the treatment of water using photocatalysis will materialise when current advancements in reactor engineering and the field of materials research are integrated. Based on the approaches that follow, researchers who want to enhance the treatment of water using photocatalysis technologies without falling into the traps that have stifled advancement over the course of the past three decades should consider the following (Loeb et al. 2018):

(i) Broaden the parameters used to evaluate the effectiveness of photocatalytic water treatment plants.

(ii) Assess the efficacy of photocatalysts under precisely defined circumstances.

(iii) Explore innovative methodologies and pursue revolutionary photocatalysts while acknowledging the difficulties of incorporating materials from other fields.
(iv) Develop and evaluate materials for targeted uses.

5.3 Phytoremediation

We can employ a variety of strategies and approaches to mitigate, eliminate, and rectify the adverse effects of pollutants discharged into the environment. Using plants to decrease the amount of contaminants in soil is an economical approach that minimises the threat to both people's well-being and the environment caused by polluted areas (Keith et al. 2024; Jeevanantham et al. 2019). The existence of noxious substances in water has an adverse effect on the aquatic environment by impeding the passage of light, hence inhibiting the process of photosynthesis in aquatic plants (Abbas et al. 2016). The plants predominantly absorb the pollutants through their root systems, thereby preventing toxicity. In addition, root systems possess a substantial surface area that collects and absorbs vital hydration and nutrition necessary for growth, as well as insignificant pollutants. This aids in wastewater purification by removing impurities and making it clean (Keith et al. 2024; Khan et al. 2022). We frequently use adsorption remedies and environmentally friendly technologies derived from plants to remove contaminants. Recent years have recorded significant advancements in this domain (Khan et al. 2022). Experiments have shown that biomass adsorbents have significant and reusable adsorption capabilities, which vary depending on the operating conditions and application methods. Parts of plants, including leaves, fruit, and other sections, effectively eliminate pollutants (Kadhom et al. 2020). This section examines plants' role in the removal of various pollutants from wastewater. Furthermore, it emphasizes the critical role that plants play in treating the most prevalent contaminants found in wastewater.

5.3.1 Diverse Phytoremediation Techniques

Jeevanantham et al. (2019) employ phytoremediation technology, a novel and environmentally friendly method, to identify, break down, and eliminate various forms of contaminants in nature's surroundings. We employ plant species to eliminate toxins that have detrimental impacts on human health and other biological systems. The processes that turn pollutants into harmless molecules are phytostabilization, rhizodegradation, phytofiltration, phytoextraction, photodegradation, phytovolatilization, and phytoaccumulation (Table 5.1) (Khan et al. 2022).

Phytostabilization employs methods of adsorption, which involve the adsorption of toxins in groundwater or dirt directly to the root or their accumulation within the rhizosphere. This prevents the migration of contaminants from one location to

Table 5.1 Diverse methodologies for phytoremediation and their mechanisms. (Reproduced from ref (Khan et al. 2022), Copyright (2022), under Creative Commons Open Access Licence from De Gruyter)

Techniques	Application	Pollutants	Mechanism	Description	Application part
Phytodesalination	Soil	Organic salts	Reducing salt by conversions	Halophytes' ability to extract salts from soils	Inside the tissues of plants
Rhizodegradation	Soil	Inorganic chemicals	Accumulation within the root zone	Rhizospheric microbial degradation of organic	Rhizosphere
Phytofiltration rhizofiltration	Water	Inorganic substances, heavy metals	Adsorption absorption	Aquatic plants absorb pollutants from polluted water sources	Aerial parts or roots
Phytodegradation phytotransformation	Soil and water	Organics	Decomposition within the root zone of plants	Chemical breakdown facilitated by plant enzymes	Inside the tissues of plants
Phytoextraction phytoaccumulation	Not often found in water or soil	Heavy metals and inorganics	Hyperaccumulation	Contaminants are taken up by the roots and then moved to higher areas.	Shoots
Phytostabilization	Water and soil	Heavy metals and inorganics	Complexation via precipitation and sorption	Roots' ability to move about in soil and access pollutants	Decrease in the rhizosphere
Phytovolatilizatio	Water and soil	Several chemical and heavy metals	Volatilization by leaves	Transformation of contaminants into a gaseous state	Discharge into the environment

another within the surrounding area. Rhizodegradation, which involves the degradation and breakdown of pollutants within the rhizosphere, transforms them into nontoxic or less hazardous molecules (Khan et al. 2022).

The process of breaking down wastes or contaminants through metabolic pathways is known as phytodegradation. During this process, plants absorb metals or garbage derived from their surroundings or sewage and break them down into harmless molecules with the assistance of several enzymes (Khan et al. 2022).

Phytovolatilization assimilates various forms of wastewater and transforms them into harmless substances. Leaves emit these harmless substances into the atmosphere through transpiration. Phytoaccumulation refers to the process of storing waste materials in different plant sections from either the surrounding ecosystem or effluent, including foliage, stems, and roots. We have employed various plant types to eliminate diverse organic wastes, heavy metals, and other pollutants (Fig. 5.4) (Khan et al. 2022).

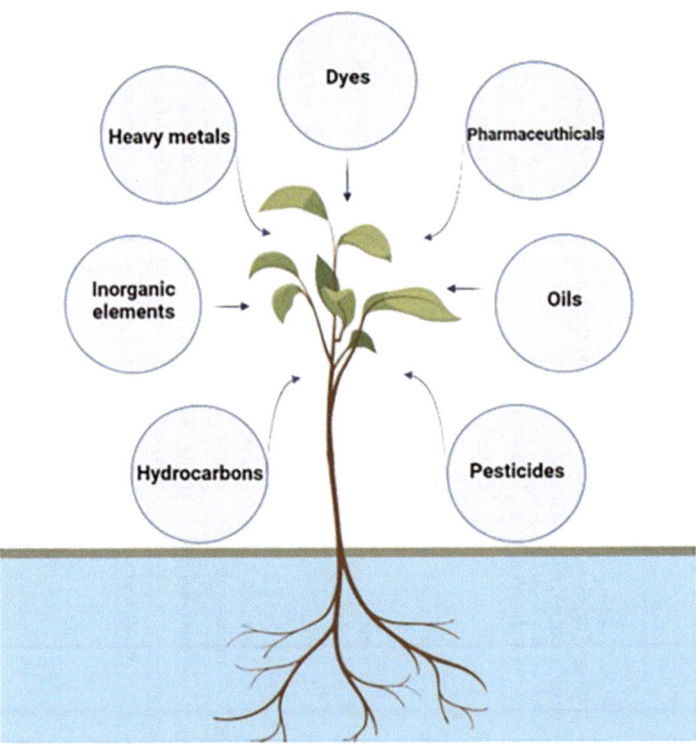

Fig. 5.4 Diverse functions of plants in the elimination of contaminants. (Reproduced from ref (Khan et al. 2022), Copyright (2022), under Creative Commons Open Access Licence from De Gruyter)

5.3.2 Phytoremediation of Wastewater Contaminants

5.3.2.1 Dye Removal

Harmful compounds in water negatively impact aquatic plants, preventing photosynthesis and affecting the water environment (Abbas et al. 2016). Various industries find the S. molesta plant highly effective in eliminating hazardous dyes (Al-Baldawi et al. 2020). Researchers commonly employ eco-friendly technologies and adsorption procedures derived from plants for color elimination, with phytoremediation emerging as a recent area of study. Biomass adsorbents have significant and reusable adsorption capabilities, depending on operating conditions and application methods. The use of botanical extracts to cure textile dyes is an emerging topic, with macrophytes being highly successful in removing chemicals like Acid Orange 7 and hazardous sulfonated anthraquinones. Several plant species, such as *Aster amellus, Glandularia pulchella, and Zinnia angustifolia,* have effectively eliminated colouring substances. Aminated sunflower seed hulls were able to retain more dye than their unmodified counterparts. Lignocellulosic wastes and pineapple plant stems have shown higher affinity for cationic dyes, while polyphenylene sulphide (PPS) has the ability to regenerate when exposed to acid, suggesting its potential as a biosorbent for eliminating BB3 (Chan et al. 2016). Four types of adsorbents are categorised as being produced from biomass: tea trash, rice ash, pineapple leaf, byproducts of the agroindustry, *Eichhornia crassipes, and Bacopa monnieri* plants. These plant components have shown exceptional efficacy in eliminating colour dyes, with enzymes found in these plant components highly recommended for industrial applications (Khan et al. 2022).

5.3.2.2 Detoxification of Heavy Metals

Heavy metals (HMs) pose significant risks to the natural world and can have detrimental impacts on human health. Researchers have devised various methodologies for the elimination of heavy metals, but they come with high costs compared to the plant-based approach for hazardous metal removal in industrial settings (Khan et al. 2022). Several research studies have investigated the efficacy of cost-effective and environmentally friendly (produced from plants) adsorbents for HMs removal (Salifu et al. 2024; Joseph et al. 2019). The global issue of HM buildup in various environmental media has arisen as a result of human operations like quarrying, urbanisation, and industrial production. Researchers are currently using plant-based approaches, specifically phytoremediation, to eliminate HMs from these effluents (Muthusaravanan et al. 2018). We are currently developing phytoremediation, an economically efficient and ecologically sustainable approach, for long-term application. Aquatic plants exhibit considerable efficacy in removing heavy metal pollutants. Duckweed (*Lemna minor*) and several plant species possess high metal accumulation efficacy (Ali et al. 2020). The utilisation of plant-based waste for the purpose of removing metals from industrial wastewater has generated considerable interest due to its cost-effectiveness and notably higher rate of removal. Several functional groups contribute to this enhanced removal rate. Coconut trash and black oak bark have the ability to extract metals, including lead, cadmium, and mercury (Alalwan et al. 2020). The utilisation of trunk fibre waste from several date palm kinds for the purpose of heavy metal removal has yielded promising outcomes. However, research is still underway to implement this project on a large scale, primarily due to cost-effectiveness and availability concerns (Muthusaravanan et al. 2018). Plants are utilised across several industries to extract heavy metals from polluted soil, restore ecosystems contaminated by metals, and mitigate ongoing detrimental impacts on organisms (Salifu et al. 2024; Muthusaravanan et al. 2018). Chemicals derived from greenery and other microbes serve as agents for HM elimination (Muthusaravanan et al. 2018). Atomic absorption spectroscopy (Nazir et al. 2020) has shown that the hyacinth (E. crassipes) plant has a lot of potential to get rid of mercury, cadmium, and arsenic.

5.3.2.3 Pesticide Elimination

Agricultural operations are primarily responsible for water contamination due to the use of pesticide compounds, which have a significant impact on the ecosystem and aquatic creatures (Khan et al. 2022). Developing cutting-edge technologies is crucial for effectively purifying tainted water and reducing pollution levels. Phytoremediation employs a four-step procedure to eliminate various forms of contamination, using plants to absorb contaminants and accumulate pollutants in their tissues for metabolization (Khan et al. 2022). This process uses transpiration to eliminate volatile organic hydrocarbons through leaf evaporation and releases exudates from a distinct plant to eliminate diverse contaminants. These exudates stimulate the growth and function of microorganisms, such as microbes and mycorrhizal fungus consortia, which can remove many types of pollutants. Various plants that grow in water, like Elodea canadensis, L. minor, and E. crassipes, are used for water pollution remediation due to their exceptional capacity to absorb contaminants, vigorous process of photosynthesis, simplicity of harvesting, and quick rates of growth (Chander et al. 2018).

Eichhornia crassipes, additionally called water hyacinths, exhibits enhanced proficiency in the absorption and breakdown of pesticides, suggesting it has the potential to be a cost-effective approach for removing pesticides from water (Rezania et al. 2015). *Lemna minor*, also known as duckweed, exhibits rapid growth within a week and demonstrates resilience to frigid temperatures (Prasertsup and Ariyakanon 2011). The rhizofiltration approach is a cost-effective way for these plants to remove organic contaminants, including pesticides and heavy metals. Dosnon-Olette et al. (2010) carried out a study using two types of duckweed plants to effectively eliminate fungicides, such as dimethomorph, from sewage from farms.

5.3.2.4 Pharmaceutical Wastes Elimination

Pharmaceutical chemicals, including ibuprofen (IBU), carbamazepine, and clofibric acid, have been found in groundwater, wastewater, and drinking water worldwide (Khan et al. 2022). Researchers such as Zhang et al. (2016) created nanoscale-built strategies that utilise wetlands light-expansive clay aggregates (LECA) and Typha spp. to study their capacity to eliminate these substances (Khan et al. 2022). The system's seasonal variability was established, showing high removal efficiency for clofibric acid, carbamazepine, and IBU throughout the summer, and 26% for clofibric acid in winter. The elimination rates were marked by an early surge in speed, with over 50% of the substance being removed within 6 h. However, larger-scale studies are needed to determine if this system can effectively treat pharmaceutical-contaminated water in wastewater treatment. Zhang et al. (2016) also created a system using four wetland plant species, *Typha, Phragmites, Iris*, and *Juncus,* to eliminate IBU and IOH from a spiking cultural solution. All plant species completely eliminated IBU within a 24-day period, while the elimination of IOH varied between 13 and 80%. *Typha* and *pragmites* demonstrated high efficacy in the removal of IOH and IBU, with a first-order rate of 0.38 and 0.06 day^{-1}.

5.3.3 Perspective

Plants have demonstrated their efficacy during the procedure for purifying wastewater, making them an essential component in future remediation efforts. Scientists can assess the plants' ability to remove pollutants from wastewater, a task that small-sized bacteria cannot accomplish. This section concisely explores various forms of phytoremediation that target abundant contaminants present in wastewater, which pose significant risks to the surroundings and human well-being. Scientists should consider combining plant-based microbes to optimise their performance, due to the fact that microbes have a long-standing track record in the breakdown and remediation of pollutants. At present, there is a lack of research on the utilisation of plants for the purpose of eliminating radioactive elements. However, exploring these concepts could prove advantageous for scientists.

5.4 Energy Generation: Microbial Fuel Cells Perspective

Microbial fuel cells (MFCs) provide eco-friendly solutions for both wastewater treatment and the production of bioelectricity, yet they do possess significant constraints. Expanding the MFC process poses a significant obstacle, as well as increasing total expenses and energy usage. Therefore, further research is necessary to develop technologies that integrate MFCs with environmental benefits and contribute to achieving sustainability goals (Sonawane et al. 2024; Nawaz et al. 2020).

Connecting wastewater treatment with energy production is crucial, and this can be achieved by utilising various reactor designs of microbial fuel cells. The section provides a comprehensive examination of MFC layout, process, various varieties of effluents, and the microbes found in effluents. Additionally, it explores the elements that influence the performance of MFCs in generating energy. Figure 5.5 illustrates

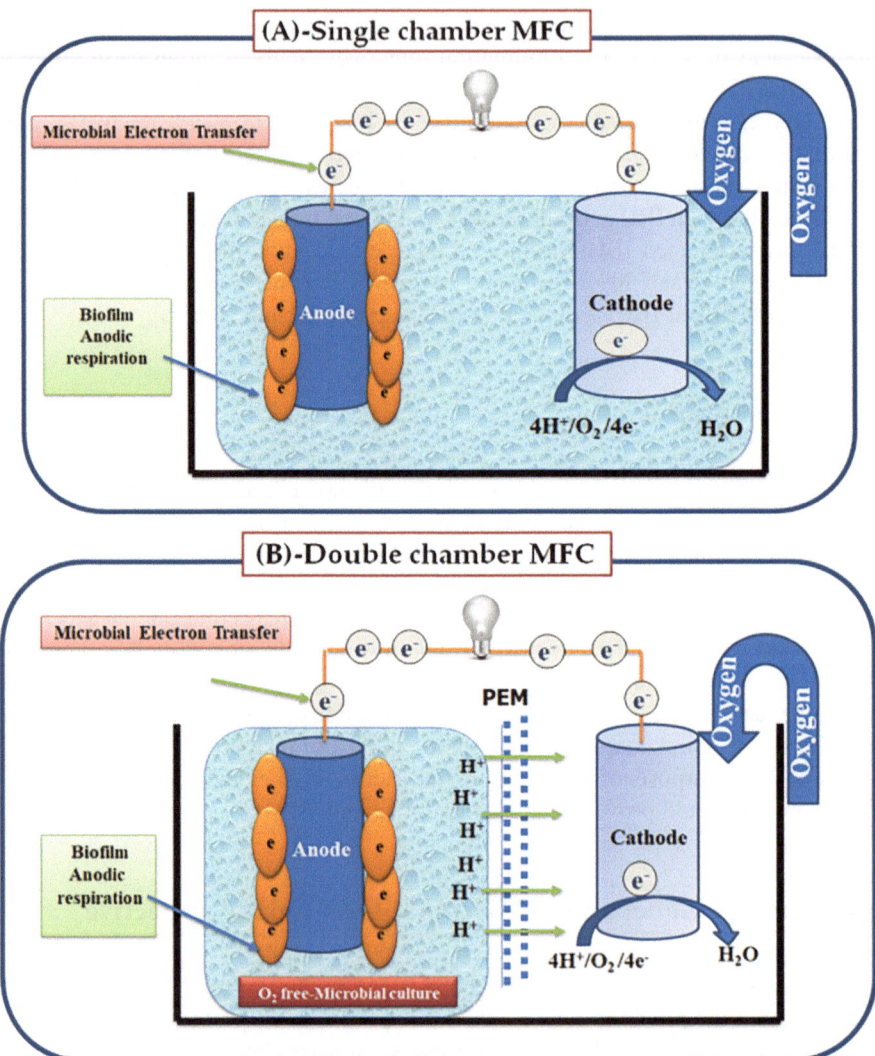

Fig. 5.5 (**a**) An arrangement of a single chamber MFC. (**b**) Design of an MFC with two chambers that are divided by a PEM membrane. (Reproduced from ref (Hassan et al. 2021), Copyright (2021), under Creative Commons Open Access Licence from MDPI)

the operational concepts of a standard MFC and how it works to produce energy (Hassan et al. 2021).

Allen's study on MFCs reveals that these devices utilize electrochemical energy conversion to produce biofilms, which serve as biocatalysts for electricity generation (Allen and Bennetto 1993). The MFC is made up of two chambers: one anodic chamber, which is maintained in an anaerobic state using nitrogen gas, and the other cathodic chamber, which is an aerobic section with an electrode for high-potential electron acceptors, often oxygen (Sonawane et al. 2024). Microbes are found in wastewater that is treated in the anodic chamber. An electrode for high-potential electron acceptors is in the aerobic cathodic chamber section. The PEM, which is made up of Nafion and Ultrex, is what separates the two chambers and works as a selective barrier in PEM fuel cells (Hassan et al. 2021).

5.4.1   Producing Bioelectricity from Wastewater Using MFC

Wastewater serves as a highly efficient means of storing renewable energy. We employ a microbial fuel cell to treat wastewater and generate power in a cost-effective manner (Sonawane et al. 2024). Wastewater, whether from animal, human, or food processing sources, is used as a substrate. A normal MFC uses the organic contaminants in wastewater and their potential electrical energy, specifically carbonaceous chemicals, to make electricity. A biochemical conversion system accomplishes this (Tatinclaux et al. 2018). Because of its capacity to utilise renewable energy sources, MFC is an optimal innovation for the generation of bioelectricity. Microbes in effluents serve as biological catalysts, facilitating the oxidation of organic or inorganic substances. They undergo metabolic processes to convert organic materials into energy and facilitate their own growth, simultaneously producing electrons, protons, and CO_2.

Electrons generated by microbial reduction gravitate towards the anode. Cytochromes, or redox-active proteins, adhere to the positively charged surface through a biological catalytic reaction. An exterior circuit, typically consisting of a Cu wire, directs electrons towards the cathode. Simultaneously, the microorganisms generate protons, which then migrate towards the cathode via the PEM. In the cathode chamber, protons will generate hydrogen gas (H_2) and hydroxyl ions, while electrons will be converted into pristine and uncontaminated water (H_2O) by a reaction with a powerful oxidising agent such as oxygen (Hamelers et al. 2010). Ferricyanide is a seldom-used oxidising agent within the cathodic compartment (Ucar et al. 2017). Two different types of metals—precious metal, specifically platinum, and non-precious metal, specifically palladium—help the process of oxygen reduction at the cathode (Nawaz et al. 2022).

5.4.2 Mechanism of Electron Transfer in MFC

Electron transfer in bacteria, such as *Geobacter spp.* and *Shewanella spp.,* involves
the transfer of electrons between bacteria and electrodes, facilitated by biomole-
cules like proteins (Kumar et al. 2018). There are two main methods for electron
transfer: Direct Electron Transfer (DET) and Mediated Electron Transfer (MET)
(Kumar et al. 2016). Figure 5.6 illustrates the schematic depiction of electron transfer pathways. In MFCs, bacteria lack proteins with the ability to undergo electrochemical reactions on their surface, necessitating the presence of electroactive
metabolites or mediators. These mediators can be synthetic or natural, depending on
the specific microorganism (Rahimnejad et al. 2015). Cells release electrons after
crossing the outer cellular lipid membrane and plasma wall, typically absorbing
them through oxygenation and the involvement of various intermediate substances.
The diminished mediator then transfers the electrons onto the anode electrode,
resulting in a negative charge. The mediator then reverts back to its initial oxidized
form, resuming the cycle. Direct electron transfer is more cost-effective and non-
toxic, making it more cost-effective. MFCs function without a mediator or external
electron carriers, and the process of transporting electrons is primarily dependent on
metal-reducing bacteria, making it crucial for processing effluent that has high concentrations of HM (Katz et al. 2003).

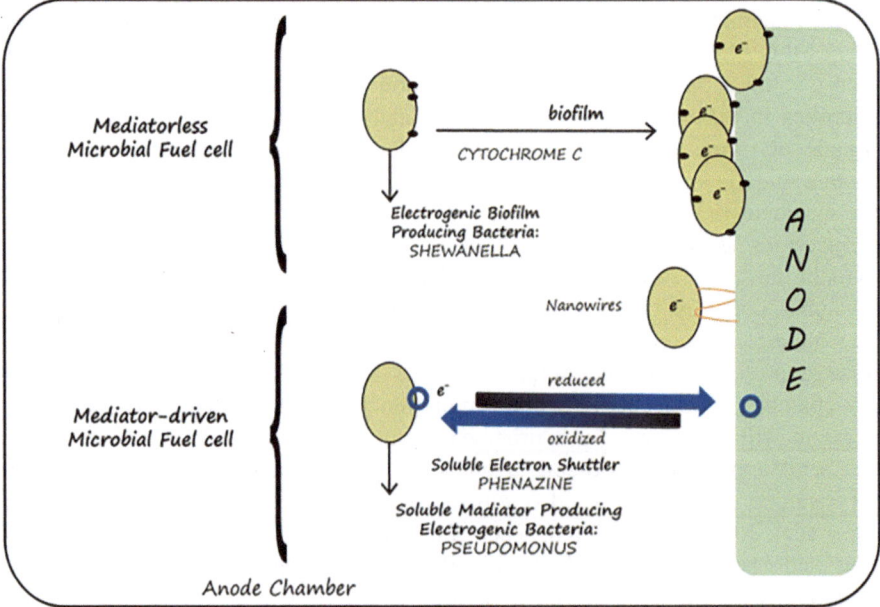

Fig. 5.6 Illustration depicting the mechanism of electron transport in an MFC. (Reproduced from
ref (Nawaz et al. 2022), Copyright (2022), under Creative Commons Open Access Licence from
Elsevier)

This process of electron transfer is necessary for energy production to work because it needs enzymes that can do electrochemical reactions, the right outside opposition, and a biofilm to form at the anode. Proper transport of protons across the membrane is essential, as improper transport can reduce electrical current, imped electron transport, and microbial activity. We prefer mediator-less electron transfer due to its enhanced safety, cost-effectiveness, and ability to generate maximum power (Gil et al. 2003).

5.4.3   Microorganisms in MFCs

Electrode-reducing and electrode-oxidising microorganisms play crucial roles in MFCs as they perform metabolic activities. Electrode-reducing microbes facilitate the reduction of electrodes and transport electrons in the direction of the anode, while electrode-oxidising microorganisms accept electrons from the cathode and catalyse the reduction of compounds like acetate and carbon dioxide. Electricigens, engaged in the process of oxidising organic substances, are responsible for electrode oxidation. The choice of microorganisms depends on their ability to transmit electrons from a substrate to an anode (Nawaz et al. 2020).

MFC utilize a variety of microorganisms for electricity generation, including pure cultures and mixed cultures (Shehab et al. 2017). MFC technology commonly uses pure cultures like Geobacter (Lovley et al. 1993) and Shewanella (Gorby et al. 2006). However, there are still numerous unidentified microbes that could significantly contribute to electricity generation. Some powerful electricigens include Archaebacteria, which can endure harsh environments; Acidobacteria, which can metabolize various substances; Cyanobacteria, which are ecologically beneficial and photosynthetic microorganisms; and Proteobacteria, which can directly transmit electrons to electrodes. Researchers investigated eukaryotes, such as yeasts, for their potential in MFCs, given their fast growth rate and lower energy production compared to bacterial strains (Nawaz et al. 2022). The genetic makeup of these strains has undergone significant advancements, increasing energy production. Exoelectrogens, such as Geobacter, Shewanella, Pseudomonas, and Clostridium, are also notable examples of microorganisms that can contribute significantly to electricity generation in MFCs. Overall, the use of pure cultures and mixed cultures in MFC technology offers promising opportunities for generating electricity from diverse microorganisms (Nawaz et al. 2022).

MFC employs microbial cultures, a combination of several types of microorganisms, to enhance biological stability. These cultures are sourced from various places like dirt, seafloor, naturally occurring microbes' communities, and brewery effluent (Munoz-Cupa et al. 2021). Wastewater, including activated sludge, household water, metal reduction agents, and waste water from a municipality, contains microbiological species capable of generating power. Researchers are investigating the use of effluent as an incubator in MFCs, where photosynthesis microbes are a viable source of electricity. Rosenbaum et al. (2005) demonstrated the environmental benefits of this approach.

5.4.4 Methods for Treating Wastewater with MFCs: Design and Configuration

Both the biocatalysts and the reactor design have an impact on the efficiency of MFCs. The essential components of a fuel cell consist of electrodes, a proton exchange membrane, and a resistor. The layout will have a direct impact on the cell's electricity generation, financial resources, and overall efficiency. Efficient design of microbial fuel cells (MFCs) necessitates consideration of several aspects, such as form, electrode type, size, and stacking direction (Gul et al. 2021). There are different varieties of MFCs, as the reactor's architecture dictates. The subsequent sections provide a comprehensive analysis of various MFC designs. We classify the MFC model as either singular or dual compartment, and we can also combine it with other models (Pant et al. 2010). Double-chamber microbial fuel cells (MFCs) are mostly utilised due to their inclusion of ion exchange membranes, two distinct compartments, and impressive power production. Despite its evolution from the double chamber, the single chamber has demonstrated more favourable outcomes (Call and Logan 2008). Figure 5.7A-F depicts the schematic illustration of various MFCs used for wastewater treatment.

5.4.5 Determinants of MFC Efficiency

The microbial fuel cell is considered a highly exciting technological advancement for the simultaneous effluent remediation and effective production of power (Bose et al. 2019). The choice of electrode and temperature, substrate loading rate, membrane material, pH, external resistance, and biofilm formation are just a few of the variables that affect MFC performance and power generation. To enhance energy output, it is necessary to optimise these elements. Figure 5.8 depicts the web diagram illustrating the various elements that influence the performance of MFCs and their ability to generate energy. The subsequent sections provide a comprehensive analysis of the impact of each individual aspect.

5.4.6 Optimising the Efficiency of MFC

MFCs are vital to energy production and wastewater treatment due to their numerous advantages, including promoting environmental sustainability. However, we have yet to develop a design that can effectively reduce exorbitant expenses and energy consumption. Double-chamber MFCs possess a clear advantage over single-chamber MFCs in terms of their superior power output and ability to efficiently remove waste water on a large scale. However, the reduced need for internal resistance in small-scale single-chambered MFCs has yielded promising results, leading to lower costs and improved efficacy. Arranging MFCs in cylindrical, horizontal, vertical, and flat-plate shapes, connected in parallel and series, improves power generation (Nawaz et al. 2022). Expanding the reactor size leads to larger volumes, which in turn reduces power and consequently causes an increase in intrinsic impedance. To address this issue, we can interconnect multiple MFCs in a series configuration to increase the total surface area. When used together, the possible power outputs increased in comparison to using separate setups or separate electrodes. Currently, in the absence of an established layout that effectively optimises both power-generating power and treating wastewater all at once, each configuration has its own advantages and disadvantages, and it is difficult to scale up a single or combined design. Future scalability of MFC requires consideration of several elements, including stacking direction, electrode material, reactor longevity, material choice, and cost efficiency (Nawaz et al. 2022).

5.4.7   What Lies Ahead?

Enhancing the efficiency of MFCs is imperative in order to achieve an optimal design. This can be achieved by successfully fulfilling a set of specific objectives. The key factors for enhancing microbial electrogenicity are (Nawaz et al. 2022): (i) augmenting the ability of microorganisms to generate electricity, (ii) carefully

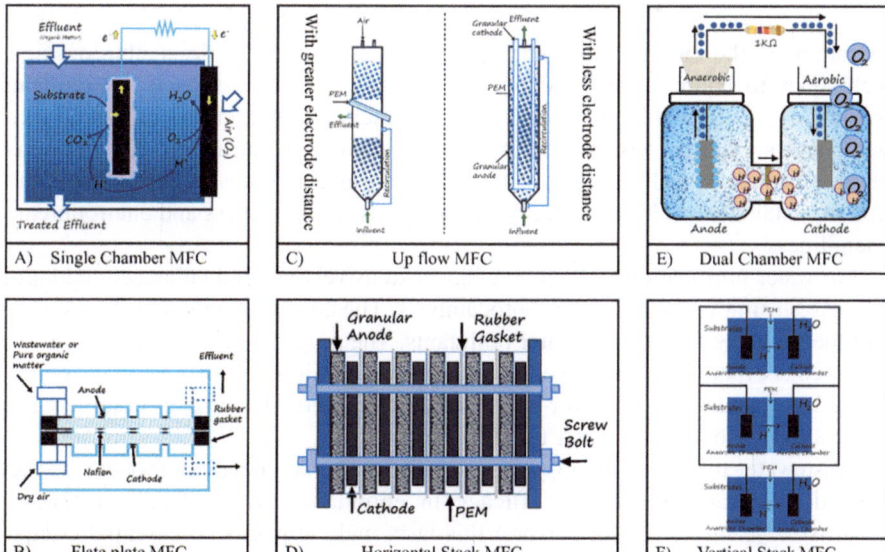

Fig. 5.7 Illustration depicting many categories of MFCs. (Reproduced from ref (Nawaz et al. 2022), Copyright (2022), under Creative Commons Open Access Licence from Elsevier)

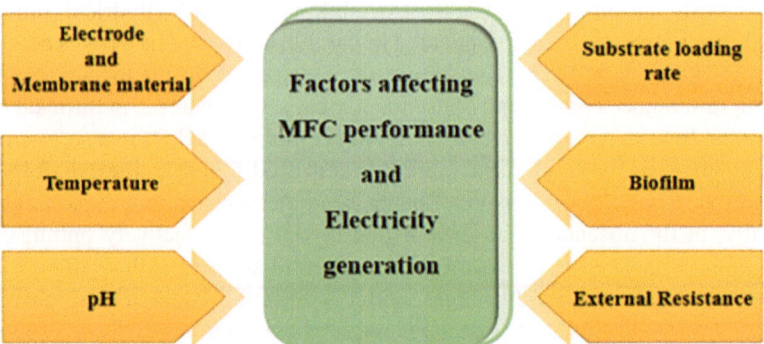

Fig. 5.8 Flowchart illustrating the parameters that determine the efficacy of a MFC in generating bioelectricity. (Reproduced from ref (Nawaz et al. 2022), Copyright (2022), under Creative Commons Open Access Licence from Elsevier)

choosing the electrode material, (iii) optimising operational parameters, (iv) effectively recovering by-products, (v) minimising capital expenses, (vi) maximising power generation, and (vii) commercialising hybrid/integrated technologies based on MFCs.

5.5   Conclusions

In summary, the emphasis has been placed on employing ecologically conscious techniques to address the issue of wastewater treatment, which is a significant worldwide concern. Green remediation techniques aim to minimize ecological harm and maximize positive environmental impacts by utilizing photocatalysis, phytoremediation, and fuel cell technologies. These techniques enhance the production of sustainable and green energy, ensuring cost-effectiveness and environmental friendliness.

For water purification, researchers have extensively studied enhanced oxidation processes using semiconductor photocatalysis. The aim is to improve reaction rates and mechanisms for treating water pollutants. The progress of photocatalytic water treatment is thoroughly analysed, evaluating its feasibility and potential applications. We suggest solutions for scientists and engineers to develop commercially applicable photocatalytic water treatment processes.

Wastewater treatment involves the use of plants as a crucial component, as they have demonstrated significant efficacy in eliminating toxins from wastewater. Small-sized bacteria cannot perform this task, making plants an essential component. Novel approaches to phytoremediation have been explored for various toxins present in wastewater, posing significant environmental and human health risks.

MFCs offer numerous advantages in energy production and wastewater treatment, promoting environmental sustainability. However, we have yet to develop a design that simultaneously optimizes power generation and wastewater treatment. Each configuration has its advantages and disadvantages, making it challenging to scale up a single or combined design. The future scalability of MFCs requires consideration of material selection, reactor durability, electrode material, cost effectiveness, and stacking direction.

References

Abbas A, Al-Amer AM, Laoui T, Al-Marri MJ, Nasser MS, Khraisheh M, Atieh MA (2016) Heavy metal removal from aqueous solution by advanced carbon nanotubes: critical review of adsorption applications. Sep Purif Technol 157:141–161. https://doi.org/10.1016/j.seppur.2015.11.039

Alalwan HA, Kadhom MA, Alminshid AH (2020) Removal of heavy metals from wastewater using agricultural byproducts. J Water Supply Res Technol AQUA 69(2):99–112. https://doi.org/10.2166/aqua.2020.133

Al-Baldawi IA, Abdullah SRS, Almansoory AF, Ismail NI, Hasan HA, Anuar N (2020) Role of Salvinia molesta in biodecolorization of methyl orange dye from water. Sci Rep 10(1):13980. https://doi.org/10.1038/s41598-020-70740-5

Ali S, Abbas Z, Rizwan M, Zaheer IE, Yavaş İ, Ünay A et al (2020) Application of floating aquatic plants in phytoremediation of heavy metals polluted water: a review. Sustain For 12(5):1927. https://doi.org/10.3390/su12051927

Allen RM, Bennetto HP (1993) Microbial fuel-cells: electricity production from carbohydrates. Appl Biochem Biotechnol 39:27–40. https://doi.org/10.1007/BF02918975

Anwer H, Mahmood A, Lee J, Kim KH, Park JW, Yip AC (2019) Photocatalysts for degradation of dyes in industrial effluents: opportunities and challenges. Nano Res 12:955–972. https://doi.org/10.1007/s12274-019-2287-0

Banerjee S, Pillai SC, Falaras P, O'shea KE, Byrne JA, Dionysiou DD (2014) New insights into the mechanism of visible light photocatalysis. J Phys Chem Lett 5(15):2543–2554. https://doi.org/10.1021/jz501030x

Benotti MJ, Stanford BD, Wert EC, Snyder SA (2009) Evaluation of a photocatalytic reactor membrane pilot system for the removal of pharmaceuticals and endocrine disrupting compounds from water. Water Res 43(6):1513–1522. https://doi.org/10.1016/j.watres.2008.12.049

Bose D, Sridharan S, Dhawan H, Vijay P, Gopinath M (2019) Biomass derived activated carbon cathode performance for sustainable power generation from Microbial Fuel Cells. Fuel 236:325–337

Call D, Logan BE, (2008) Hydrogen production in a single chamber microbial electrolysis cell lacking a membrane. Environ Sci Technol 42(9):3401–3406

Chan SL, Tan YP, Abdullah AH, Ong ST (2016) Equilibrium, kinetic and thermodynamic studies of a new potential biosorbent for the removal of basic blue 3 and Congo red dyes: pineapple (Ananas comosus) plant stem. J Taiwan Inst Chem Eng 61:306–315. https://doi.org/10.1016/j.jtice.2016.01.010

Chander PD, Fai CM, Kin CM (2018, June) Removal of pesticides using aquatic plants in water resources: a review. IOP Conf Ser Earth Environ Sci 164(1):012027. https://doi.org/10.1088/1755-1315/164/1/012027

Chen X, Liu L, Yu PY, Mao SS (2011) Increasing solar absorption for photocatalysis with black hydrogenated titanium dioxide nanocrystals. Science 331(6018):746–750. https://doi.org/10.1126/science.1200448

Chen J, Loeb S, Kim JH (2017) LED revolution: fundamentals and prospects for UV disinfection applications. Environ Sci Water Res Technol 3(2):188–202. https://doi.org/10.1039/C6EW00241B

Choi W, Termin A, Hoffmann MR (2002) The role of metal ion dopants in quantum-sized TiO2: correlation between photoreactivity and charge carrier recombination dynamics. J Phys Chem 98(51):13669–13679. https://doi.org/10.1021/j100102a038

Dosnon-Olette R, Couderchet M, El Arfaoui A, Sayen S, Eullaffroy P (2010) Influence of initial pesticide concentrations and plant population density on dimethomorph toxicity and removal by two duckweed species. Sci Total Environ 408(10):2254–2259. https://doi.org/10.1016/j.scitotenv.2010.01.057

Elsalamony RA (2016) Advances in photo-catalytic materials for environmental applications. Res Rev J Mater Sci 4(2):26–50

Fujishima A, Honda K (1972) Electrochemical photolysis of water at a semiconductor electrode. Nature 238(5358):37–38. https://doi.org/10.1038/238037a0

Gil GC, Chang IS, Kim BH, Kim M, Jang JK, Park HS, Kim HJ (2003) Operational parameters affecting the performannce of a mediator-less microbial fuel cell. Biosens Bioelectron 18(4):327–334. https://doi.org/10.1016/S0956-5663(02)00110-0

Gorby YA, Yanina S, McLean JS, Rosso KM, Moyles D, Dohnalkova A et al (2006) Electrically conductive bacterial nanowires produced by Shewanella oneidensis strain MR-1 and other microorganisms. Proc Natl Acad Sci 103(30):11358–11363. https://doi.org/10.1073/pnas.0604517103

Grčić I, Puma GL (2017) Six-flux absorption-scattering models for photocatalysis under wide-spectrum irradiation sources in annular and flat reactors using catalysts with different optical properties. Appl Catal B Environ 211:222–234. https://doi.org/10.1016/j.apcatb.2017.04.014

Gul S, Balkhi MH, Abubakr A, Shah TH, Bhat BA, Bhat FA, Javeed H (2021) A study on physicochemical parameters of Manasbal Lake, Kashmir, India. J Pharm Innov 10(9):554–559

Hamelers HV, Ter Heijne A, Sleutels TH, Jeremiasse AW, Strik DP, Buisman CJ (2010) New applications and performance of bioelectrochemical systems. Appl Microbiol Biotechnol 85:1673–1685. https://doi.org/10.1007/s00253-009-2357-1

Hassan RY, Febbraio F, Andreescu S (2021) Microbial electrochemical systems: principles, construction and biosensing applications. Sensors 21(4):1279. https://doi.org/10.3390/s21041279

Hoffmann MR, Martin ST, Choi W, Bahnemann DW (1995) Environmental applications of semiconductor photocatalysis. Chem Rev 95(1):69–96. https://doi.org/10.1021/cr00033a004

Hurum DC, Agrios AG, Gray KA, Rajh T, Thurnauer MC (2003) Explaining the enhanced photocatalytic activity of Degussa P25 mixed-phase TiO2 using EPR. J Phys Chem B 107(19):4545–4549. https://doi.org/10.1021/jp0273934

Islam M, Kumar S, Saxena N, Nafees A (2023) Photocatalytic degradation of dyes present in industrial effluents: a review. ChemistrySelect 8(26):e202301048. https://doi.org/10.1002/slct.202301048

Jeevanantham S, Saravanan A, Hemavathy RV, Kumar PS, Yaashikaa PR, Yuvaraj D (2019) Removal of toxic pollutants from water environment by phytoremediation: a survey on application and future prospects. Environ Technol Innov 13:264–276. https://doi.org/10.1016/j.eti.2018.12.007

Joseph L, Jun BM, Flora JR, Park CM, Yoon Y (2019) Removal of heavy metals from water sources in the developing world using low-cost materials: a review. Chemosphere 229:142–159. https://doi.org/10.1016/j.chemosphere.2019.04.198

Kadhom M, Albayati N, Alalwan H, Al-Furaiji M (2020) Removal of dyes by agricultural waste. Sustain Chem Pharm 16:100259. https://doi.org/10.1016/j.scp.2020.100259

Katz E, Shipway AN, Willner I (2003) Biochemical fuel cells. In: Vielstich W, Gasteiger HA, Lamm A (eds) Handbook of fuel cells–fundamentals, technology and applications, 1. Wiley, pp 355–381

Katz A, McDonagh A, Tijing L, Shon HK (2015) Fouling and inactivation of titanium dioxide-based photocatalytic systems. Crit Rev Environ Sci Technol 45(17):1880–1915. https://doi.org/10.1080/10643389.2014.1000763

Keith BF, Lam EJ, Montofré ÍL, Zetola V, Bech J (2024) The scientific landscape of phytoremediation of tailings: a bibliometric and scientometric analysis. Int J Phytoremediation 8:1–19. https://doi.org/10.1080/15226514.2024.2373427

Khan A, Balakrishnan K, Katona T (2008) Ultraviolet light-emitting diodes based on group three nitrides. Nat Photonics 2(2):77–84. https://doi.org/10.1038/nphoton.2007.293

Khan MM, Adil SF, Al-Mayouf A (2015) Metal oxides as photocatalysts. J Saudi Chem Soc 19(5):462–464. https://doi.org/10.1016/j.jscs.2015.04.003

Khan AU, Khan AN, Waris A, Ilyas M, Zamel D (2022) Phytoremediation of pollutants from wastewater: a concise review. Open Life Sci 17(1):488–496. https://doi.org/10.1515/biol-2022-0056

Kim J, Lee CW, Choi W (2010) Platinized WO3 as an environmental photocatalyst that generates OH radicals under visible light. Environ Sci Technol 44(17):6849–6854. https://doi.org/10.1021/es101981r

Kumar R, Singh L, Zularisam AW (2016) Exoelectrogens: recent advances in molecular drivers involved in extracellular electron transfer and strategies used to improve it for microbial fuel cell applications. Renew Sust Energ Rev 56:1322–1336. https://doi.org/10.1016/j.rser.2015.12.029

Kumar R, Singh L, Zularisam AW, Hai FI (2018) Microbial fuel cell is emerging as a versatile technology: a review on its possible applications, challenges and strategies to improve the performances. Int J Energy Res 42(2):369–394. https://doi.org/10.1002/er.3780

Liu J, Liu Y, Liu N, Han Y, Zhang X, Huang H et al (2015) Metal-free efficient photocatalyst for stable visible water splitting via a two-electron pathway. Science 347(6225):970–974. https://doi.org/10.1126/science.aaa3145

Loeb S, Hofmann R, Kim JH (2016) Beyond the pipeline: assessing the efficiency limits of advanced technologies for solar water disinfection. Environ Sci Technol Lett 3(3):73–80. https://doi.org/10.1021/acs.estlett.6b00023

Loeb SK, Alvarez PJ, Brame JA, Cates EL, Choi W, Crittenden J et al (2018) The technology horizon for photocatalytic water treatment: sunrise or sunset? Environ Sci Technol 53(6):2937–2947. https://doi.org/10.1021/acs.est.8b05041

Lovley DR, Giovannoni SJ, White DC, Champine JE, Phillips EJP, Gorby YA, Goodwin S (1993) Geobacter metallireducens gen. nov. sp. nov., a microorganism capable of coupling the complete oxidation of organic compounds to the reduction of iron and other metals. Arch Microbiol 159:336–344. https://doi.org/10.1007/BF00290916

Malini B, Gandhimathi R (2024) An overview of photocatalytic degradation of agricultural pollutants in water. Int J Green Energy 21(12):2843–2857. https://doi.org/10.1080/15435075.2024.2332918

Matafonova G, Batoev V (2018) Recent advances in application of UV light-emitting diodes for degrading organic pollutants in water through advanced oxidation processes: a review. Water Res 132:177–189. https://doi.org/10.1016/j.watres.2017.12.079

Mierzwa JC, Rodrigues R, Teixeira AC (2018) UV-hydrogen peroxide processes. In: Advanced oxidation processes for waste water treatment. Academic, pp 13–48. https://doi.org/10.1016/B978-0-12-810499-6.00002-4

Munoz-Cupa C, Hu Y, Xu C, Bassi A (2021) An overview of microbial fuel cell usage in wastewater treatment, resource recovery and energy production. Sci Total Environ 754:142429. https://doi.org/10.1016/j.scitotenv.2020.142429

Muthusaravanan S, Sivarajasekar N, Vivek JS, Paramasivan T, Naushad M, Prakashmaran J et al (2018) Phytoremediation of heavy metals: mechanisms, methods and enhancements. Environ Chem Lett 16:1339–1359. https://doi.org/10.1007/s10311-018-0762-3

Nawaz A, Hafeez A, Abbas SZ, Haq IU, Mukhtar H, Rafatullah M (2020) A state-of-the-art review on electron transfer mechanisms, characteristics, applications and recent advancements in microbial fuel cells technology. Green Chem Lett Rev 13(4):365–381. https://doi.org/10.1080/17518253.2020.1854871

Nawaz A, ul Haq I, Qaisar K, Gunes B, Raja SI, Mohyuddin K, Amin H (2022) Microbial fuel cells: insight into simultaneous wastewater treatment and bioelectricity generation. Process Saf Environ Prot 161:357–373. https://doi.org/10.1016/j.psep.2022.03.039

Nazir MI, Idrees I, Idrees P, Ahmad S, Ali Q, Malik A (2020) Potential of water hyacinth (Eichhornia crassipes L.) for phytoremediation of heavy metals from waste water. Biol Clin Sci Res J 2020(1):e006. https://doi.org/10.54112/bcsrj.v2020i1.6

Nedjimi B (2021) Phytoremediation: a sustainable environmental technology for heavy metals decontamination. SN Appl Sci 3(3):286. https://doi.org/10.1007/s42452-021-04301-4

Okamoto KI, Yamamoto Y, Tanaka H, Tanaka M, Itaya A (1985) Heterogeneous photocatalytic decomposition of phenol over TiO2 powder. Bull Chem Soc Jpn 58(7):2015–2022. https://doi.org/10.1246/bcsj.58.2015

Pan C, Zhu Y (2015) A review of BiPO4, a highly efficient oxyacid-type photocatalyst, used for environmental applications. Cat Sci Technol 5(6):3071–3083. https://doi.org/10.1039/C5CY00202H

Pant D, Van Bogaert G, Diels L, Vanbroekhoven K (2010) A review of the substrates used in microbial fuel cells (MFCs) for sustainable energy production. Bioresour Technol 101(6):1533–1543.

Prasertsup P, Ariyakanon N (2011) Removal of chlorpyrifos by water lettuce (Pistia stratiotes L.) and duckweed (Lemna minor L.). Int J Phytoremediation 13(4):383–395. https://doi.org/10.1080/15226514.2010.495145

Puma GL, Brucato A (2007) Dimensionless analysis of slurry photocatalytic reactors using two-flux and six-flux radiation absorption–scattering models. Catal Today 122(1–2):78–90. https://doi.org/10.1016/j.cattod.2007.01.027

Rahimnejad M, Adhami A, Darvari S, Zirepour A, Oh SE (2015) Microbial fuel cell as new technology for bioelectricity generation: a review. Alex Eng J 54(3):745–756. https://doi.org/10.1016/j.aej.2015.03.031

Rezania S, Ponraj M, Din MFM, Songip AR, Sairan FM, Chelliapan S (2015) The diverse applications of water hyacinth with main focus on sustainable energy and production for new era: an overview. Renew Sust Energ Rev 41:943–954. https://doi.org/10.1016/j.rser.2014.09.006

Rosenbaum M, Schröder U, Scholz F (2005) In situ electrooxidation of photobiological hydrogen in a photobioelectrochemical fuel cell based on Rhodobacter sphaeroides. Environ Sci Technol 39(16):6328–6333. https://doi.org/10.1021/es0505447

Salifu M, John MA, Abubakar M, Bankole IA, Ajayi ND, Amusan O (2024) Phytoremediation strategies for heavy metal contamination: a review on sustainable approach for environmental restoration. J Environ Prot 15(4):450–474. https://doi.org/10.4236/jep.2024.154026

Sathishkumar P (2024) Current trends and future perspectives. In: Photocatalysis for 'energy and environmental applications. Springer, Singapore. https://doi.org/10.1007/978-981-97-1939-6

Shehab NA, Ortiz-Medina JF, Katuri KP, Hari AR, Amy G, Logan BE, Saikaly PE (2017) Enrichment of extremophilic exoelectrogens in microbial electrolysis cells using Red Sea brine pools as inocula. Bioresour Technol 239:82–86. https://doi.org/10.1016/j.biortech.2017.04.122

Sonawane AV, Rikame S, Sonawane SH, Gaikwad M, Bhanvase B, Sonawane SS et al (2024) A review of microbial fuel cell and its diversification in the development of green energy technology. Chemosphere 350:141127. https://doi.org/10.1016/j.chemosphere.2024.141127

Suri RP, Liu J, Hand DW, Crittenden JC, Perram DL, Mullins ME (1993) Heterogeneous photocatalytic oxidation of hazardous organic contaminants in water. Water Environ Res 65(5):665–673. https://doi.org/10.2175/WER.65.5.9

Tatinclaux M, Gregoire K, Leininger A, Biffinger JC, Tender L, Ramirez M et al (2018) Electricity generation from wastewater using a floating air cathode microbial fuel cell. Water-Energy Nexus 1(2):97–103. https://doi.org/10.1016/j.wen.2018.09.001

Tsukamoto D, Shiraishi Y, Sugano Y, Ichikawa S, Tanaka S, Hirai T (2012) Gold nanoparticles located at the interface of anatase/rutile TiO2 particles as active plasmonic photocatalysts for aerobic oxidation. J Am Chem Soc 134(14):6309–6315. https://doi.org/10.1021/ja2120647

Ucar D, Zhang Y, Angelidaki I (2017) An overview of electron acceptors in microbial fuel cells. Front Microbiol 8:643. https://doi.org/10.3389/fmicb.2017.00643

Van Gerven T, Mul G, Moulijn J, Stankiewicz A (2007) A review of intensification of photocatalytic processes. Chem Eng Process Process Intensif 46(9):781–789. https://doi.org/10.1016/j.cep.2007.05.012

Wang W, Tade MO, Shao Z (2015) Research progress of perovskite materials in photocatalysis- and photovoltaics-related energy conversion and environmental treatment. Chem Soc Rev 44(15):5371–5408. https://doi.org/10.1039/C5CS00113G

Xiong Z, Ma J, Ng WJ, Waite TD, Zhao XS (2011) Silver-modified mesoporous TiO2 photocatalyst for water purification. Water Res 45(5):2095–2103. https://doi.org/10.1016/j.watres.2010.12.019

Zhang Y, Tang ZR, Fu X, Xu YJ (2010) TiO2– graphene nanocomposites for gas-phase photocatalytic degradation of volatile aromatic pollutant: is TiO2– graphene truly different from other TiO2– carbon composite materials? ACS Nano 4(12):7303–7314. https://doi.org/10.1021/nn1024219

Zhang Y, Lv T, Carvalho PN, Arias CA, Chen Z, Brix H (2016) Removal of the pharmaceuticals ibuprofen and iohexol by four wetland plant species in hydroponic culture: plant uptake and microbial degradation. Environ Sci Pollut Res 23:2890–2898. https://doi.org/10.1007/s11356-015-5552-x

Zhao C, Pelaez M, Dionysiou DD, Pillai SC, Byrne JA, O'Shea KE (2014) UV and visible light activated TiO2 photocatalysis of 6-hydroxymethyl uracil, a model compound for the potent cyanotoxin cylindrospermopsin. Catal Today 224:70–76. https://doi.org/10.1016/j.cattod.2013.09.042

Chapter 6
Future Threats and Their Remediation Processes

Abstract The American Water Works Association (AWWA) emphasizes the importance of sustainable water management, a crucial requirement for the survival of all living organisms. The growing population and urbanization have intensified the demand for reliable water supplies, posing challenges in ensuring sufficient water supply and effective wastewater treatment. The Sustainable Development Goals emphasize the need for sustainable wastewater management, advocating for the safe use of purified drinking water and accessible treatment procedures. The AWWA has identified five crucial problems affecting progress towards a sustainable and adaptable water future: sustainability, technology, economics, governance, and social and demographic concerns. To make sure that our methods for cleaning wastewater are both cost-effective and safe for the environment, we use photocatalysis, phytoremediation, microbial fuel cells, activated biochar, electrocoagulation, and biomass-based hydrogels.

Keywords Sustainability · Eco-friendly · Cost-effective · Photocatalysis · Phytoremediation

6.1 Future Challenges or Threats

Environmental degradation is a critical issue that is causing irreparable harm to the planet. Urbanization and technological advancements have deteriorated crucial environmental components, such as air, water, soil, noise, and light. Industrial waste materials, including plastics, heavy metals, nitrates, fossil fuel combustion, acid rain, oil spills, and industrial toxins, contribute to this degradation. These pollutants disrupt ecological equilibrium and pose a threat to extinction for various animal and avian species. Acid rain's degradation of vegetation disrupts animals' natural habitats. To effectively address environmental health challenges, it is essential to understand the connection between environmental contaminants and mental well-being.

N. Saxena et al., *Water Pollution and Remediation*, SpringerBriefs in Water Science and Technology, https://doi.org/10.1007/978-3-031-76301-4_6

Prioritizing planning and implementation is crucial to achieving the intended outcomes in this field. Addressing these issues is essential to protecting human health and the environment (Vinod et al. 2023).

The American Water Works Association (AWWA) brings together the diverse water community to advance public health, safety, the economy, the water workforce, and the environment. It provides education to individuals working in the field of water and promotes the importance of secure and environmentally friendly water practices. AWWA highlights the crucial role of water in local economies and the environment, emphasizing the need for effective management and distribution of water resources (AWWA-2050-Social/Demographics 2023). The report emphasizes the need for safety, affordability, and community well-being. Water-related issues will impact economies, population distribution, and technology development over the next three decades. The AWWA recognizes five key challenges: sustainability, technology, economics, governance, and social and demographic factors. Addressing these challenges will shape future efforts related to wastewater treatment techniques, ensuring the sustainability of local economies and the well-being of living organisms.

6.1.1   Sustainability

Effectively overseeing our world's finite water resources and constructed infrastructure for water is of utmost importance. Climate change poses significant concerns. It will result in more intense and unpredictable weather, such as prolonged periods of drought and heatwaves, a heightened frequency of hurricanes and wildfires, and severe winter storms. In the future, it will be necessary to have skilled and creative management of our most important natural resource, along with innovative methods to maintain strong and resilient water infrastructure (AWWA-2050-Sustainability 2022).

6.1.2   Technology

In the midst of the fourth industrial revolution, water-related professionals now have the ability to utilise innovative technologies that are revolutionising their interactions with the resources of water, water networks, and the communities they represent. The advancements in data, analytics, the Internet of Things (IoT), machine learning, and artificial intelligence will gradually provide customers with more control and impact the functioning of water systems. The implementation of new technology will effectively address intricate issues while occasionally creating unforeseen difficulties (AWWA-2050- Technology 2022).

6.1.3   Economics

Water plays a crucial role in driving economic growth in North American cities and worldwide. The water community is increasingly facing the challenge of enhancing efficiency and effectiveness, while simultaneously addressing the growing demands for infrastructure improvements. It is crucial to take into account significant economic aspects such as regionalization, supply chain resilience, decentralised treatment, ESG methods for risk and value assessment, and the advantages of a circular economy. Rate-setting will take place in a world that is well aware of the difficulties related to fairness and economics (AWWA-2050- Economics 2023).

6.1.4   Governance

The functions of federal, provincial, state, and local governments have a substantial influence on the operation and regulation of water utilities. Governance and economics will both shape the future paradigm of water utilities. Certain towns may choose regional solutions in order to achieve greater efficiencies. As regulatory frameworks progress, communities will need to assess novel strategies, such as tailored standards and distributed treatment (AWWA-2050- Governance 2023).

6.1.5   Social/Demographics

The growing public interest and concern regarding water quality and equity necessitates the collective effort of all communities to enhance public confidence. Concurrently, the possibility of population movements between urban and rural regions is giving rise to difficulties in managing resources and infrastructure, as well as necessitating the implementation of water solutions driven by local communities. To address population expansion in water-stressed areas, creative strategies will be required to effectively manage scarce water resources (AWWA-2050- Social/ Demographics 2023).

6.2   Methods of Green Remediation for Treating Wastewater

Water is a vital necessity for the survival of living organisms worldwide, and the increasing population and urbanisation have heightened the need for secure and accessible natural reserves. The combination of explosive industrialization, global warming, and a huge human population has led to a substantial challenge in ensuring enough water supply and effectively managing wastewater. Addressing these difficulties is crucial for sustainable water resources and wastewater management. The Sustainable Development Goals worldwide prioritise sustainable wastewater management, promoting the safe use of purified drinking water, and ensuring accessible and sufficient treatment methods are available. The primary goal is to purify the potable water supply by eliminating toxins, organic poisons, and heavy metals via environmentally friendly technologies and management strategies (Dhak et al. 2023).

Green remediation strategies assess the ecological impact of water treatment and explore innovative methods to minimize damage and maximize ecosystem benefits. These methods improve sustainable energy generation, ensuring cost efficiency and environmental compatibility. Photocatalysis, phytoremediation, and fuel cell technologies, particularly MFCs, are crucial in combating water pollution.

Photocatalysis is an efficient method for removing contaminants from water, particularly heavy metals, which are detrimental to ecosystems and human well-being. Advanced oxidation processes (AOPs) efficiently decompose organic waste into full mineralization, producing reactive oxygen species (ROS) (Islam et al. 2024). Ozone and hydrogen peroxide are chemical oxidants used in conventional AOPs. UV therapy frequently serves as an energy source to generate ROS, which oxidizes and removes water pollutants (Loeb et al. 2018). People widely recognize photocatalytic water filtration as more efficient than homogeneous-phase advanced AOPs. Phytoremediation is a cost-effective and environmentally benign remediation approach that shows promise as a preventive tool against water contamination, particularly heavy metals (Islam et al. 2024). Understanding the mechanisms underlying pollutants such as heavy metal accumulation and plants' ability to withstand them is crucial for enhancing the efficiency of phytoremediation.

6.2.1 Photocatalysis for Green Remediation

In photocatalysis, photocatalysts capture light energy and eliminate toxic contaminants from wastewater. Access to uncontaminated drinking water is an essential entitlement for all individuals; nevertheless, the issue of water pollution persists as a significant worldwide problem. Conventional techniques of treating wastewater frequently do not effectively deal with newly identified toxins and long-lasting pollutants (Sakshi 2024).

Water treatment widely uses photocatalysis for various purposes, including the elimination of organic pollutants, disinfection of microorganisms, and the breakdown of complex contaminants. Photocatalytic oxidation is a highly effective process for breaking down and eliminating organic dyes, medicines, antibiotics, insecticides, and industrial pollutants. Additionally, it results in the conversion of contaminants into non-toxic substances by mineralization. Moreover, photocatalytic disinfection provides a way of sterilizing water that does not include the use of chemicals, reducing the potential hazards associated with traditional disinfection techniques such as chlorination (Sakshi 2024). Photocatalysis has numerous benefits in water treatment (Sakshi 2024):

 (i) Photocatalysis enables the total breakdown of contaminants while minimizing the need for chemical substances.
 (ii) It demonstrates a wide range of effectiveness against different types of pollutants.
(iii) This procedure is straightforward, easy, reusable, and economical.
(iv) The contemporary development of biodegradable clay-based photocatalysts has bestowed upon them an additional environmental benefit.

However, obstacles such as the need for appropriate light conditions, reduced efficacy, limited lifespan, and instability hinder the broad deployment of photocatalysts. Addressing these issues is critical in order to fully utilize the promise of photocatalysis in water treatment. We can address these issues by developing innovative photocatalytic materials, designing new reactors, and implementing efficient operating procedures (Sakshi 2024).

6.2.2 Phytoremediation for Green Remediation

Plants play a crucial role in soil remediation, reducing the harmful impacts of pollutants on the environment. They absorb pollutants through their root systems, which absorb essential nutrients and contaminants. People often use adsorption therapies and eco-friendly plant-derived technologies to eliminate pollutants. Recent advancements in this field have shown that biomass adsorbents have significant and repeatable adsorption capacities. Plants, including leaves and fruit, have the ability to efficiently remove contaminants from wastewater. This highlights the importance of plants in treating common pollutants in wastewater.

The following are the primary benefits of phytoremediation (Islam et al. 2024):

 (i) Environmentally responsible and enduringly viable.
 (ii) Economically feasible
(iii) Addresses a wide range of pollutants

Phytoremediation is a promising technique for removing contaminants, although it does have several downsides, including (Kafle et al. 2022, Islam et al. 2024):

 (i) The process of restoration will require a significant amount of time.
 (ii) Research on phytoremediation is limited and controlled, necessitating extensive field experiments to fully understand its potential.
(iii) Limited effectiveness of metal hyperaccumulators and poor growth rate and insufficient biomass.
(iv) Having extensive roots and a large root biomass will enhance the effectiveness of the cleanup process. Therefore, it is necessary to arrange plants in a staggered manner in order to optimize their phytoremediation potential.
 (v) The potential risk of toxicity in animals and other organisms due to the consumption of plants with a high concentration of pollutants can be mitigated through effective phytoremediation measures, including treatment, removal, or monitoring.

6.3   Alternative Techniques and Their Illustrations

Liu et al. (2022) successfully eliminated tetracycline from an aqueous solution by activating biochar from penicillin fermentation waste with K_2CO_3. An IKBCH, a type of microporous biochar with a sufficiently large specific surface area, was produced using the impregnated method with hydrochloric acid ageing, resulting in excellent adsorption capabilities for tetracycline. With 1 gramme of microporous biochar per litre of water, which equals 200 milligrammes per litre, a 99.91% clearance rate of tetracycline (TC) was reached. How adsorption works primarily involves electrostatic adsorption, π-π interaction, hydrogen bonding, pore filling, and hydrogen bonds. The IKBCH process of adsorbing TC demonstrated exceptional stability, ensuring that there was no secondary contamination in the environment. Furthermore, IKBCH demonstrated excellent adsorption capabilities against a variety of contaminants, making it a highly effective adsorbent with great potential.

Gradinac et al. (2022) pioneered the use of titanium electrodes to purify water in steel manufacturing factories. The electrocoagulation (EC) method assessed the effectiveness of scale ion removal in water using titanium rod electrodes. The study examined prototype electrodes. In a closed system, the study focused on demonstrating the efficacy and efficiency of using titanium electrodes to remove hardness from processed and makeup water. This was achieved by utilising a Universal Environmental Technologies system (UET). Researchers thought they could fix the issue by improving the UET system, making sure the water was balanced, and adding a self-regulating system to keep an eye on things like alkalinity, total hardness, corrosion, magnesium, calcium, iron, and chlorine, making sure they didn't go over their limits. The use of autoregulation technology in a closed system can lead to improved efficiency and high-quality water.

Zhang et al. (2022) have described the application of biomass-based hydrogels for the efficient removal of toxic heavy metals from aqueous solutions. This study involved the preparation of a hydrogel with excellent adsorption properties. They formed the hydrogel by combining porous tobacco straw with STS-PAA, a polyacrylic acid. Using UV light, they pretreated the tobacco straw and then incorporated acrylic acid and potassium acrylate into a polymer material. They conducted an investigation to examine the adsorption performance of metal ions. They conducted an investigation to examine the impact of various periods of adsorption, temperature, and pH levels, as well as the impact of metal ion levels on the quantity of heavy metal ions adsorbed. The hydrogel exhibited a significant capacity for removing Pb^{2+}, Hg^{2+}, and Cd^{2+} ions from an aqueous solution. The adsorption of Pb^{2+} was highly efficient. When the concentration started at 4.0 mmol/L and the pH was 6, it reached equilibrium. The amounts of Pb^{2+}, Hg^{2+}, and Cd^{2+} that were absorbed were 1.49 mmol g^{-1}, 1.02 mmol L^{-1}, and 0.94 mmol g^{-1}, respectively. In a pseudo-first-order model, the kinetic results revealed that Pb^{2+}, Hg^{2+}, and Cd^{2+} adhere to the framework, indicating the adsorption of all three of these heavy metal ions through physical means in a monolayer. An investigation in thermodynamics determined that the adsorption of Pb^{2+}, Hg^{2+}, and Cd^{2+} onto the STS-PAA is characterised by an endothermic ($\Delta H > 0$), entropy increase ($\Delta S > 0$), and non-spontaneous reaction.

6.4   Conclusion

There are five primary challenges that will significantly impact progress towards a sustainable and flexible water future. The issues encompass sustainability, technology, economics, governance, as well as social and demographic factors. These challenges will significantly impact and shape all future efforts related to wastewater treatment methods, requiring the adoption of a comprehensive and integrated strategy. The green remediation strategies aim to minimize ecological harm and optimize positive environmental impacts by employing techniques such as photocatalysis and phytoremediation. These approaches enhance the production of renewable and environmentally friendly water resources, ensuring both cost-effectiveness and environmental compatibility. In recent times, researchers have also employed activated biochar, electrocoagulation, and biomass-based hydrogels, among other methods.

References

Dhak D, Chiavola A, Mishra A, Dhak P (2023) Innovations and challenges in green and sustainable water purification and waste water management. Front Chem 11:1235757. https://doi.org/10.3389/fchem.2023.1235757

Gradinac J, Jovović A (2022) Investigation regarding the application of the titanium electrode for the water treatment plant in a steel manufacturing plant. Front Chem 10:1065332. https://doi.org/10.3389/fchem.2022.1065332

Islam MM, Saxena N, Sharma D (2024) Phytoremediation as a green and sustainable prospective method for heavy metal contamination: a review. RSC Sustain 2:1269–1288. https://doi.org/10.1039/D3SU00440F

Kafle A, Timilsina A, Gautam A, Adhikari K, Bhattarai A, Aryal N (2022) Phytoremediation: mechanisms, plant selection and enhancement by natural and synthetic agents. Environ Adv 8(2022):100203. https://doi.org/10.1016/j.envadv.2022.100203

Liu Y, Gao W, Yin S, Liu R, Li Z (2022) Efficient removal of tetracycline from aqueous solution by K2CO3 activated penicillin fermentation residue biochar. Front Chem 10:1078877. https://doi.org/10.3389/fchem.2022.1078877

Loeb SK, Alvarez PJ, Brame JA, Cates EL, Choi W, Crittenden J et al (2018) The technology horizon for photocatalytic water treatment: sunrise or sunset? Environ Sci Technol 53(6):2937–2947. https://doi.org/10.1021/acs.est.8b05041

Sakshi KM (2024, April 28) Publishing associate: researcher and writer at save the water™ l. https://savethewater.org/photocatalysis-in-wastewater-treatment-harnessing-the-power-of-light/

Vinod KG, Anoop Y, Chandra M, Sushma Y, Neeraj K (2023) Green chemistry approaches to environmental sustainability: status, challenges and prospective. Elsevier. https://doi.org/10.1016/C2022-0-00289-0

Water 2050, American water works association, Economics think tank, (2023, January 23–25) The Penn Club, New York, NY

Water 2050, American water works association, Governance think tank, (2023, February 27–March 1) Reservoir Center for Water Solutions, Washington, D.C.

Water 2050, American water works association, Social/Demographics think tank, (2023, April 26–28) Birmingham Civil Rights Institute, Birmingham

Water 2050, American water works association, Sustainability think tank, (2022, September 21–23) Springs Preserve, Las Vegas

Water 2050, American water works association, Technology think tank, (2022, December 5–7) Computer History Museum, Mountain View

Zhang M, Zhou Y, Wang F, Chen Z, Zhao X, Duan W et al (2022) Preparation of biomass-based hydrogels and their efficient heavy metal removal from aqueous solution. Front Chem 10:1054286. https://doi.org/10.3389/fchem.2022.1054286

Conclusion

Water contamination is a significant issue in both urban and rural areas. This pollution is primarily due to fertilizing farmland, unregulated disposal of bio-wastes, and excessive livestock breeding in rural areas. The health impacts of water pollution are significant, as the presence of pollutants makes it unsuitable for ingestion, posing a public health concern. Researchers have linked birth deformities, methemoglobinemia, typhoid, dysentery, cholera, and infectious hepatitis to consuming contaminated drinking water.

In today's world, water is a considered as a vital resource for local economies and their natural environments. The current imperative is to efficiently manage and allocate water supplies, prioritise safety and affordability, and foster community well-being. In the coming 30 years, water-related concerns will have a significant impact on economies, population distribution, and technological advancement. As a result, the use of international watercourses for purposes other than navigation following the Industrial Revolution had a significant impact on the development of international water law.

The convergence of rapid industrialization, climate change-induced global warming, and a burgeoning human population has presented a significant predicament in guaranteeing an adequate water supply and efficiently handling wastewater. It is essential to tackle these challenges in order to achieve sustainable management of water resources and wastewater. Globally, the Sustainable Development Goals emphasise the importance of sustainable wastewater management, advocating for the safe use of purified drinking water, and guaranteeing the availability of accessible and adequate treatment procedures.

The examination of classic pollutants such as heavy metals and nutrients has been carried out; yet, the development of approaches that are both efficient and inexpensive for the analysis of pollutants continues to be an issue. The fast growth of human society has resulted in the poisoning of water, which has negatively impacted not just the economy but also public health and the environment. Emerging pollutants pose significant risks to both human individuals and the ecosystems in

N. Saxena et al., *Water Pollution and Remediation*, SpringerBriefs in Water Science and Technology, https://doi.org/10.1007/978-3-031-76301-4

which they are found. When attempting to ascertain the level of water pollution, it is essential to conduct an assessment of the water quality by means of the Water Quality Index (WQI). For the purpose of quantifying water quality, the Water Quality Index (WQI) is a basic method that takes into consideration a certain parameter and provides a single value. The contamination of groundwater, the pollution of microbiological systems, the pollution of nutrients, the pollution of surface water, and the pollution of chemical substances all have a significant influence on a significant portion of the water resources on Earth. This is a significant cause for worry because of the growing need for residential, industrial, and agricultural applications combined. The physical, chemical, and biological aspects of water are taken into consideration while determining the water's quality.

Technologies such as photocatalysis, phytoremediation, and fuel cells are utilised in order to address the issue of wastewater treatment. These approaches are environmentally conscious efforts. In addition to ensuring cost-effectiveness and environmental friendliness, these strategies improve the production of energy that is both sustainable and environmentally friendly. There has been a significant amount of research conducted on photocatalytic water treatment, and novel approaches to phytoremediation have been investigated for the treatment of a variety of toxins that are found in wastewater.

There are five primary obstacles that must be overcome in order to achieve a water future that is both sustainable and adaptable. These obstacles are sustainability, technology, economics, governance, and social and demographic circumstances. For the purpose of attaining a water future that is both sustainable and adaptive, it is vital to implement green remediation strategies that concentrate on minimising ecological damage and optimising the positive effects on the environment through the use of technologies such as photocatalysis, phytoremediation, and fuel cell technology.